The State of American
Hot Rodding

The State of American Hot Rodding

Interviews on the Craft and the Road Ahead

DAVID LAWRENCE MILLER

McFarland & Company, Inc., Publishers
Jefferson, North Carolina

Library of Congress Cataloguing-in-Publication Data

Names: Miller, David L., 1948– author.
Title: The state of American hot rodding : interviews on the craft and the road ahead / David Lawrence Miller.
Description: Jefferson, North Carolina ; McFarland & Company, Inc., Publishers, 2018 | Includes bibliographical references and index.
Identifiers: LCCN 2018008685 | ISBN 9781476672915 (softcover : acid free paper) ♾
Subjects: LCSH: Hot rods—United States. | Hot rods—United States—Interviews. | Automobile racing—United States. | LCGFT: Interviews.
Classification: LCC TL236.3 .M55 2018 | DDC 629.228/60973—dc23
LC record available at https://lccn.loc.gov/2018008685

British Library cataloguing data are available

**ISBN (print) 978-1-4766-7291-5
ISBN (ebook) 978-1-4766-3181-3**

© 2018 David Lawrence Miller. All rights reserved

No part of this book may be reproduced or transmitted in any form or by any means, electronic or mechanical, including photocopying or recording, or by any information storage and retrieval system, without permission in writing from the publisher.

Front cover: Bo Huff's 1929 Ford Model A Roadster, an exhibit at his Sunnyvale, Utah, museum, in July 2010 (photograph by the author)

Printed in the United States of America

*McFarland & Company, Inc., Publishers
Box 611, Jefferson, North Carolina 28640
www.mcfarlandpub.com*

For Pete Chapouris, Tony Farnetti, Sr., Gary "Chopit" Fioto,
Blackie Gejeian, Bo Huff, Gary Meadors, and "Speedy" Bill Smith.
Rest in peace my friends.

And to the members of my "brain trust"—
Joe Alig, Gary Buehler, Eline Haukenes, Hans Jacobson
and Nick Miller-Jacobson, Renee K. Gadoua, Greg Hrehovcsik,
Roger Jetter, and Bill Goodwill.

Also to Dave and Dick Stoker,
along with the rest of the members of the 7 Valley
Street Rods Club of Cortland, New York.

Author's Note

Politeness and a general respect for you, the reader, require some context and a disclaimer. I am a geographer; my aim has always been to make sense of changing landscapes. In 44 years of driving, I have never had a speeding ticket and have never been involved in a serious accident. I am basically a law-abiding guy. People I've never met tell me that I look like Dale Earnhardt, with or without the sunglasses. I have always liked '58 Chevy Apache trucks, the sound of a big V8, and Jimmy Buffett's music. In my best moments, I am in most respects a libertarian. I am no lover of authoritarianism of any stripe. Power corrupts. So let's see if I can continue to avoid a speeding ticket in my purpose-built highway cruiser. It's time for a road trip.

Here's the disclaimer: I alone am responsible for any deviations from the facts or from what passes as current wisdom.

Table of Contents

Author's Note	vi

PART ONE: THE RUN UP

1. Introduction: Hot Rodding as Defiance of the Contemporary	3
2. My Credentials: The Dismantled Omnibus	8
3. My '58 Build Story, or How I was Adopted by a Chevy Apache Behind a South Dakota Barn	13
4. "Drive What You Build" Lessons	20

PART TWO: THE RUN OUT

5. Introduction to the Interviews	25
6. Bobby Alloway (Louisville, TN)	28
7. Bo Huff (Sunnyside, UT)	35
8. George Ross (Thompson's Station, TN)	48
9. Lee Osborne (Penn Yan, NY)	61
10. Darryl Starbird (Afton, OK)	73
11. Brylen Brajkovich (Jonestown, PA)	85
12. John Davis (Alamosa, CO)	93
13. Bob Austin (Rochester, NY)	101
14. Rick Treworgy (Punta Gorda, FL)	110
15. Rod Petty (Moab, UT)	118
16. George Roetman (Vermillion, SD)	127
17. Rick Love (San Antonio, TX)	132
18. Walt Johnson (Thorntown, IN) and Ken Holdaway (Fairview Heights, IL)	139

19. Jerry Dixey (Austintown, OH) — 153
20. Gene Winfield (Mojave, CA) — 160

PART THREE: THE RUN DOWN

21. The Road Ahead: Higher Gas Prices vs. America's Love Affair with Man Toys — 165
22. The Takeaway: Thoughts on the Future of Hot Rodding — 171
23. Street Rodding 2.0: How We Roll Forward — 175
24. Final Thoughts — 178
25. At the End, 2017 — 181

Chapter Notes — 189
Selected Bibliography — 195
Index — 197

By the time you get close to the answers, it's nearly all over.
—Merle Haggard (April 6, 1937–April 6, 2016)

Part One

The Run Up

1

Introduction: Hot Rodding as Defiance of the Contemporary

The word hot rodding has become used to describe enjoying anything modified for personal, performance, cosmetic or "daring-to-be different" reasons and has become an umbrella term covering all things modified or created.—Tom Vogele, StreetScene[1]

Hot rodders like old cars and trucks, beer, and having a good time. But who really are hot rodders? Where do gearheads come from? What do they value and why? How can this non-conformity persist in light of the aging out of baby boomers, changing technology, and the ever-increasing regimentation of everyday life? These are the questions this book focuses on. For the past several years, I have driven across America in my home-built 1958 Chevy Apache pickup, seeking answers. I have interviewed gearheads at car shows, during road tours, at shops, and in home garages. Many of my interviews were arranged, but many were serendipitous—occasioned by chance encounters when I pulled into a gas station or stopped at a restaurant or bar. Car people would come over to talk while I pumped gas or come into a diner looking for the owner of the truck parked outside. I have recorded many of these conversations.

This book presents the best of the many stories I gathered, with an eye toward preserving detail about one element of the contemporary car culture landscape before it forever changes. My effort is divided into three parts. The first (The Run Up) is an introduction and context section where I lay out the landscape and relate details about my own immersion in the gearhead world. I do this to establish my credentials, but also to illustrate what is involved in the creation of a street rod or custom. The second part (The Run Out) consists of stories from interviews with builders and collectors. I have been fortunate to interview some of the best-known names in the rod shop industry, but I also visited home workshop-based rod builders and just hung out, cruised, drank coffee, or enjoyed an occasional beer with hot rodders from all over the country. The third part (The Run Down) focuses on the question "What does it all mean?" Here I share my thoughts about the value and viability of the enterprise. It goes without saying that almost everything, from a personal transportation standpoint, is going to change. And that rust never sleeps. Although rodding culture persists, in my darker moments I think that the gearhead gene may be clipped from our DNA by "progress." I hope this is not the case.

Whatever the road ahead, hot rodding is today a vibrant American subculture. This book celebrates the persistence of rodding culture as a revolt against the regimentation of everyday life. From an evolutionary perspective, gearheads are racing against the currents of technology and the commonplace. And yet gearheads persist. They design, build, run, fix, and rebuild their rides. The road ahead may include an expiration date on their activities, but as long as there are open roads, hot rodders will defy the trend toward mechanical cluelessness, anonymous styling, driverless cars, and a "get back in line and wait to be served" mentality that permeates our current transportation landscape.

What exactly is a hot rod? For the purposes of this book, I am rolling with *StreetScene* magazine Editor Tom Vogele's definition: a vehicle that has been "modified for personal, performance, cosmetic, or daring-to-be-different reasons." And what's a gearhead? This is a person, most often male, who is totally into cars or trucks and knows—or will figure out—how to fix or modify them for better performance, appearance, and—in the case of rat rods—shock value.

Who are hot rodders? Most rodders are members of the baby boomer cohort. These guys were kids when their fathers and uncles joined the culture upon their return from World War II. In interview after interview, I heard that most of today's hot rodders became gearheads when an older family member or friend took an interest in them and taught them wrenching, triggering the expression of their gearhead gene.

Hot rod culture really began to spread after men returned from the war. They'd seen enough; they just wanted to settle in and have a good life. But for many of them the good life wasn't good enough. Maybe they'd flown B-24s in the Pacific, like my dad. Or served as a Marine at Guadalcanal like my Uncle Lawrence, or been among the 73,000 who landed on Normandy beaches on D-Day. Wrenching on an old car provided a socially acceptable, solitary opportunity to focus on something tangible.

Given what we now know about PTSD, I believe it was a form of therapy for many. The outlaw street racer nature of rodding may have offered many of them the shot of adrenalin that they craved as a tonic for the mind-numbing normalcy of everyday life. The unregimented element of the activity must have held great attraction. Maybe it was the only way to keep moving forward. As time passed, they found similar souls and formed car clubs. As their kids or neighborhood kids showed some interest, they took them under their wing. These kids, now graying baby boomers, represent the bulk of today's hot rodders.

The baby boomer cohort will continue to play an important role in the activity. This will extend perhaps through 2029, roughly a dozen years from now, when all of the boomers will be over 65. Beyond that date, because of this demographic transition and for a host of other reasons, hot rodding will be morphing into another form that only modestly reflects today's car culture landscape.

How big is hot rodding? For the most part, the activity is a cultural phenomenon that rides along quietly, under the radar of most Americans. Still, this element of car culture has its moments. Every July, just up the highway from where I live, the Syracuse Nationals attracts some 78,000 spectators and nearly 7,000 street rods and custom cruisers to a weekend show. And down Interstate 81, the Carlisle Pennsylvania Fairgrounds hosts a series of shows and swap meets that draw annually more than 500,000 participants. Large shows are organized by national organizations such as the Right Coast Association ("Cool Cars and Hot Happenings"), the Goodguys Rod & Custom Association ("Biggest and Best Hot Rodding Association in the USA"), and the National Street Rod Association ("Fun with Cars").

In 2015, at 24 Goodguys events, some 76,500 cars were entered and 1,350,000 spectators attended. And at two of the largest hot rod shows in the nation—the National Street Rod Association's Nationals held in Louisville, Kentucky and the Minnesota Street Rod Association's "Back to the 50s"—more than 10,000 rods, customs, muscle cars, and street machines assemble at each venue. Nearly 100,000 spectators attend each event. On a smaller scale, local car clubs in the Central New York region hold nearly 50 cruise-ins or shows a year. But, like I said, most of this happens under the radar of most Americans.

Is hot rodding a hobby? I cringe a bit at that question because it diminishes the significance the activity holds for hot rodders. But for lack of a better descriptor, and for the purposes of this book, hot rodding is a hobby. Of course, many individuals find employment in the industry; but most gearheads hot rod their vehicles for pleasure when they are not working. Or they get into the hobby in a big way when they retire. But hot rodders are genuinely passionate about what they do. And, as I'll describe in the pages that follow, there are inherently rebellious and non-conformist aspects to the activity.

The most useful essay I have read about the nature of hobbies was written by the ecologist Aldo Leopold in his book *Round River Journals*. It may seem odd to start a book on hot rodding with a quote from an ecologist. Most hot rodders don't consider themselves environmentalists, except perhaps in the sense that they recycle old metal. They give no premium to minimizing fuel consumption, and burnouts are the order of the day at many gatherings. But bear with me. In the book's first essay, "A Man's Leisure Time," Leopold states that his hobby of making longbows and hunting with them "for better or worse," equates "with the need of doing something queer." He writes,

> At first blush I am tempted to conclude that a satisfactory hobby must be in large degree useless, inefficient, laborious, or irrelevant. Certainly many of our most satisfying avocations today consist of making something by hand which machines can usually make more quickly and cheaply, and sometimes better.... A hobby is a defiance of the contemporary. It is an assertion of those permanent values which the momentary eddies of social evolution have contravened or overlooked.... It is an axiom that no hobby should either seek or need rational justification. To wish to do it is reason enough. To find reasons why it is useful or beneficial converts it at once to the ignominious category of an 'exercise' undertaken for health, power, or profit. Lifting dumbbells is not a hobby.

Although some hot rodders may feel a compelling urge to rationalize their hobby, from an environmental standpoint the activity is neither useful nor beneficial. Rodding is not sustainable, at least as I understand long-term prospects. Toward the end of his essay, Leopold amends his definition of a hobby to something that "entails either making something or making the tools to make it with, and then using it to accomplish some needless thing." In essence, a hobby represents a defiance of the contemporary.

Most hot rodders will probably resist the "hobbyist" label and the claim that they build needless things. I certainly do. Car guys build and modify their rides to look, run, and sound good. Hobbyists or not, if we reflect deeply enough, most gearheads can ride with Leopold's "defiance" notion. And they can get on board with his observation that "a good hobby must also be a gamble." Along this line, he ends his essay focusing on longbow making and the possibility that his effort will burst "into impotent splinters" resulting in

> another laborious month of evenings at the bench. The possible debacle is, in short, an essential element in all hobbies, and stands in bold contradistinction to the humdrum certainty that the endless belt will eventuate in a Ford.

Every hot rodder I know has orchestrated some debacle, buildwise. And every rodder laments the cookie cutter nature of today's automobile manufacturers. As one gearhead said, "You used to be able to identify a car at night by looking at its headlights in your rearview mirror." Back in the day, cars were distinctive and had style. Recently, GM produced a series of posters focusing on what car guys would consider the neutering of today's cars and trucks. One poster featured a close-up of the iconic front end of a red '57 Chevy with the tagline, "Proof that your parents were actually cool once."

In the '50s and '60s, a savvy car guy could "option out" his car in a way to create at least a semi-unique vehicle. Today, with few exceptions, you can tell a vehicle's make only by getting close enough in the daylight to see its badging. Any car guy will tell you that unique design by today's car makers has been rendered sterile. Ironically, marketers are attempting to use this as a selling point. A current Chevy TV ad shows a vehicle without badging and asks a group of hipsters to guess its make. Members of the group suggest BMW, Camry, and so forth. Then the group learns the car is a Chevy Malibu, with a starting price about $22,000.

So "looks just like" is now being marketed as a virtue. The humdrum is being drummed into us. Car guys get this; most would be quick to suggest that maybe GM should just give up the pretense of a distinct brand and sell the Malibu without any badging. Even today's off-the-rack muscle car reincarnations (Mustang or Camaro) have limited cachet among gearheads. Although these models perform well and can be tricked out with aftermarket goodies, most hot rodders want to work with old metal. They do this, Leopold would say, in defiance of the contemporary. From the standpoint of the gearhead subculture, wrenching and building your own ride is a proper way to establish independence and shape a unique identity.

If you are still with me, the last paragraph of Leopold's essay should resonate like a lumpy cam:

> A good hobby may be a solitary revolt against the commonplace or it may be the joint conspiracy of a congenial group.... In either event it is a rebellion, and if a hopeless one, all the better.... Nonconformity is the highest evolutionary attainment of social animals.... A hobby is perhaps creation's first denial of the "peck-order" that burdens the gregarious universe, and of which the majority of mankind is still a part.

I like the idea of hot rodding as a "revolt against the commonplace." This statement sits right. Ask a gearhead, and he will communicate the notion that car guys run in the countercurrents of conventional society. Hot rodders build, modify, repair, and maintain cars. They don't simply have someone at a dealership plug in a new part. When things stop, rodders figure out a way to make them go. They are outliers who have managed to persist. At one or many levels, car guys are outlaws; they defy the contemporary.

How can this small rebellion persist? It seems inevitable that driverless electric cars will push aside our fossil-fuel rides. We're about to be swamped by this next wave of change as technology eliminates driving.[2] As one writer puts it, albeit darkly, we are approaching the death throes of the hand-driven vehicle:

> Laziness and fear have formed a rare partnership, hiring technology to eliminate driving. Lane drift alarms, blind spot monitors, adaptive cruise control have been at this for years, and the self-driving car is at our gates. Soon Skynet-backed robots will be ferrying their meatbag owners to work in endless, anonymous road trains, a dense clog in the colon of the American highway.[3]

In this vision of the future, the "defiance of the contemporary" by today's rodders will not long be overlooked. Street rods, custom cars, classic trucks, and muscle cars—for reasons of pollution, lack of fuel economy, and a failure to fit in—should be the next to go, following the hoof prints left by the passing of the horse and buggy.

Whether you buy this vision or not, entropy—the gradual slide to deterioration and breakdown—will get our metal in the end. Most of us are graybeards; there is no clear next-gen with respect to taking up the wrench. And, as gearheads know well, rust never sleeps. No application of corrosion-sealing technologies will forever protect our cherished rides. Fire, landslides, sinkholes, earthquakes, tidal waves, civil unrest, economic downslides, displacement by driverless cars, laws supporting green-based transport technologies, or our aging bodies will eventually whittle the focus of our passion to relatively few vehicles in museums. We may be able to buy some time with Computer Numerical Control (CNC) machines and 3D printers, but the die seems cast, literally and figuratively.[4]

Nevertheless, virtually every hot rodder would argue that this is somewhere far down the road.[5] Rodding is a vibrant cultural phenomenon. Go to any show, and you will find people who are passionate about their ride and know every nut and bolt holding it together. They built it and they know how to fix it. They are American gearheads, and this book celebrates how they roll.

2

My Credentials: The Dismantled Omnibus

Rust Never Sleeps.—Slogan developed for Rust-Oleum paint in the 1970s by Mark Mothersbaugh and Jerry Casale, later of the band Devo[1]

As I begin this book, it is January 2008 in Upstate New York. My 1958 Chevy Apache pickup rests cold in my unheated garage. In the final stages of her build into a highway cruiser, her bed sides and tailgate are off and away getting some final bodywork, then paint by my friend Rick Busutti. The current plan is that I'll have the bed back sometime next month.

So she sits, old metal combined with new technology—an old skin over new mechanicals—missing her bedsides, cooling her custom aluminum wheels, in a sort of hibernation. When things warm up, say mid May, I'll be working on the cab; replacing door hinges and making final adjustments to the cab's doors and fenders. Then she goes back to Rick's for final work on the cab, a careful block sanding, and paint.

I intend to drive my home-built '58 across America, on a road trip in search of stories about people who build hot rods and custom cars. I'll focus on the "drive what you build" element of hot rodding. I will interview mostly folks who build "drivers," not "trailer queens" (high-end custom show vehicles).

I am embarking on this journey for several reasons. First, to my knowledge, there has been little systematic coverage of this enterprise by academics. Second, and more importantly, this is a subculture on the bubble. Hot rods and custom cars, "old metal" powered by big V8 engines, burn prodigious amounts of fossil-based fuel. As I write this [in 2008], some authorities suggest that we are at or beyond peak oil, and that what remains will become increasingly expensive. The price of a barrel of oil is well over $100 a barrel, and pump gas prices seem on an ever upward trajectory. We seem to be at some sort of a tipping point, where the viability of hot rodding is increasingly strangled by high fuel prices and undercut by concerns about environmental impacts. Are we approaching the end of an era?

As I ponder that likelihood, an oil painting at the Milwaukee Art Museum haunts me. Eastman Johnson's 1871 work, *The Old Stagecoach*, shows a group of children playing on an abandoned stagecoach. It sits in the weeds, without wheels, with four children acting the part of a team of horses in harness, others aboard the faded red coach clearly imagining their vehicle racing toward some destination. The museum's collection notes relate that Johnson painted this at the height of his career and that

> sentimental genre scenes of rustic youth and rural life were extremely popular in the wake of the Civil War, evoking a nostalgia for simpler times.... Images of innocent and carefree children offered promise for a new beginning to a generation troubled by industrialization and decaying urban conditions ... the painting is a joyful celebration of the hope of a nation as embodied in the laughter of its youth.[2]

I am haunted by this painting because, every time I look at it, I imagine an old hot rod in place of the stagecoach cab. It, too, rests in the weeds on its rails, wheels long gone, driver's door missing, windows busted out, and sheet metal rusting away. In my nightmare, children play on the hot rod's derelict shell. This is not a "joyful celebration of hope" for those of us who love old metal, yet it may be an appropriate visual metaphor for the future of rod and custom builders.

Even today this image continues to dog me; I have to tell myself, "Down, boy." In researching this picture, I located a 1907 *New York Times* article in which Johnson's paintings are called "Old School," and this painting of "boys romping over the body of a dismantled omnibus" is said to be from an era when "pictures with an amusing or merely a gently facetious meaning were readily taken by people who liked to have cheerful things about them in their homes."[3] In this case, the cheerful image is obsolete, characterized by antiquated fashion.

So like the painting's obsolete stagecoach, are today's hot rods destined for extinction? To put it bluntly, are we at the end of an era? The 2007 federal energy law requires

Eastman Johnson (American, 1824–1906). *The Old Stagecoach*, 1871. Oil on canvas. 36¼ × 60⅛ in. (92.08 × 152.72 cm). Layton Art Collection, Inc., Gift of Frederick Layton, at the Milwaukee Art Museum, L1888.22. Photographer credit: John R. Glembin.

auto manufacturers to increase fleet fuel-efficiency from 25 mpg to 35 mpg by 2020—a 40 percent increase. How long until super-sized engines in street rods are called out for their lack of efficiency and emissions? Is the rod and custom enterprise at, or approaching, the end of its viability because of increasing fuel prices, concerns about environmental impacts, and a general unraveling of an unsustainable economy? Are our custom rides "cheerful things" that cannot be supported: old metal with technology that is out of step with fuel efficiency mandates, subject to ever-increasing fuel costs and taxes, and incompatible with alternative fuels, hybrid technology and autonomous vehicles. Are we the last hurrah of an antiquated, no longer affordable, and obsolete fashion destined for the weeds at the side of the road?

There is, of course, some irony here. Hot rod and custom builders take pride in having dragged the basis of their ride from the weeds, from behind a barn, or out of a junkyard. This is so because it is already possible to assemble a "classic" vehicle such as an early '50s Chevy truck or a hot rod such as a '32 Ford with 100 percent newly manufactured sheet metal and a high-tech chassis package with a suspension that will yield the handling of a new Corvette. Last year a new metal prototype was displayed of a big-window cab for a '55–'59 Chevy truck. It is an exact copy—weld for weld—of my truck's cab, except that I have the small rear-window version.[4]

But I digress. At the turn of the last century, Johnson's paintings were called "Old School," which the dictionary defines as "having the character, manner, or opinions of a past age." The term means something quite different to members of the hot rod and street cruiser culture. Today, "Old School" refers to builders who use traditional techniques, technology, material, and finishes. This term is generally understood as a label of respect, recognizing a high level of craftsmanship bordering on the cunning. Generally speaking, a great deal of respect exists for Old School rides.

Can the elements and ethics of Old School be retooled into a new school? Dramatic changes are in the offing, with respect to fuels, technology, and materials. Which elements of custom culture will be lost and which can be maintained? How far will old metal carry into the future? What is affordable, what will be abandoned, and what will be replaced?

Looking down the road, it seems clear that our transportation will need to be based on more sustainable technologies. Will the cost of these new modes overtake our ability to afford fossil fuel-based old metal cruisers? In addition to ever-increasing costs to operate their hot rods, fewer and fewer builders will be able to assemble rides using old metal, traditional materials, and V8 engines. And whenever the value of metal for scrap recycling increases, there will be less and less original material available for customs. What's left will become increasingly expensive. As John Gilbert, editor of *Custom Classic Trucks* magazine observes:

> In the States, our raw materials are being dragged from the fields and sold for scrap due to the current high prices of scrap metal and the Europeans, Aussies, and whoever else are here in droves buying our old stuff because the dollar is weak.[5]

Thoughtful members of the custom culture have to be troubled by the picture emerging. By nature, many of the "drive-what-you-build" folks are traditionalists in the sense that they do not entirely embrace innovation. Builders are individualists, but some generalizations apply. Old metal is given the most currency, in both senses of the phrase.

Purists frown on, but tolerate, "reproduction" or newly minted "old" parts. Old metal salvaged from a scrap yard or from the weeds of the back 40, or an original stock (OS) part from the back shelf of a longtime dealership, is the best option. Flathead engines from the '30s to late '40s are highly prized, followed by Y-block Ford engines and 283–350 cubic inch GM engines running up through the muscle car motors of the '70s.

As time passes, however, there is less and less original material to work with. Steep price increases are the rule. Some substitutions are being made. For example, rides powered by new technology—fuel-injected engines such as GM's LS series, which are computerized—increasingly appear under the hoods of hot rods and cruisers at custom shows. These engines are considerably more efficient than the carbureted Chevy small block that is under the hood of many rides. Some traditionalists frown at these plug and play LS-series engines, but others have a somewhat more open attitude, suggesting, "It's OK for someone else, but I wouldn't have one under my hood."

Newly minted glass (Fiberglass) roadster and cruiser bodies are available, such as Coast to Coast's '39 Ford body, which offers a menu of options, as alternatives to the process of crafting a custom body. Most of these are incorporated into high-end rides with numerous billet aluminum and chrome accessories that many traditionalists regard as over-the-top. Others, however, respect the fact that many of these rides are well-engineered and reliable. These creations are indeed capable of being driven from coast to coast.

It's hard to know what to make of all this. Nevertheless, experience teaches us that things change and rust never sleeps. Rodding and custom culture, from this perspective, seems on the bubble. Or at the precipice. Or already gone over the edge of the cliff. Maybe hot rodders are just diehards.

Despite the signs, I cannot believe that the craftsmanship required to create, and the passion necessary to maintain, custom rides will simply fade away. I do not believe this culture will abandon its enterprise and join the rest of America at the mall or online or in self-driving cars. I'd like to believe the scarcity of materials, high fuel costs, and our dwindling discretionary income will cause us to modify, rather than abandon, our enterprise.

I'd also like to believe that the freedom of the open road will remain available, that future generations will have the means and ability to build custom vehicles for the open highway. The value of a long-distance road trip should not be underestimated. As Livy, the Roman historian of some 2,000 years ago, said, "We fear things in proportion to our ignorance of them."[6] This is a very big, very diverse, country. Increasingly, we are retreating into gated communities and gated minds. If we lose the will to move about freely and explore the country, we will be impoverished in every sense. And at a time when authoritarianism is on the rise, as individual rights are being systematically eroded by a frantic waving of the terror towel, and as freedoms are exchanged for imagined security, the notion of what can be learned on the open road in a purpose-built cruiser seems a perfectly reasonable antidote to the drift away from personal liberties.

So that's the preamble. As I stated near the beginning of this chapter, this book explores the landscape of a culture that seems at a tipping point; fuel costs and concerns about environmental impacts may strangle the enterprise. Looking far down the road, our rides may be solar and hydrogen hybrids. Will custom culture carry forward or will it be abandoned, like stagecoaches of the past? Will the next generation's children play on the hulks of rusting custom rides or will they learn the skills necessary to maintain and adapt their vehicles to keep them on the road?

These are very big questions. And you don't just walk up to most gearheads and get conversation, much less answers. For the most part, it's a closed fraternity. While you can walk up to a car displayed at a show and admire the ride, if you are a spectator, you are generally limited to descriptive information from the owner. That's about it.

Many hot rodders are reluctant to speak in depth about what they do and why they do it. They build things with their hands, not with their words. Their ride is their statement. Gearheads have their own vocabulary. In their own way, many hot rodders are thoughtful and articulate. But it's a different culture, a different tradition. You need to know the language to gather intelligence. As you can imagine, a typical college professor has little or nothing in common with the group.

Unless the college professor has the gearhead gene. Which I do. As a kid, I worked my way up from a broom in a garage. Somehow I got on an academic track, but I never lost the compulsion to repair and fix things. So, late in my career, I have built a custom 1958 Chevy pickup to carry me across America in search of answers. My truck is intended to serve, in the most literal sense, as a vehicle for conversations. She will be my calling card for interviews with hot rodders, and muscle and classic car owners.

*   *   *

In hindsight, as I work on the final edits of this book in spring 2017, my initial concerns about the impact of high fuel prices on our enterprise would seem to have been overblown. Gas prices have tanked due to a fracking-related supply boom. I would point out, however, that the promise of cheap gas has never materialized for most rodders because their traditional engines have to run with 91 octane ethanol-free fuel for which they pay a stiff premium.

Because I powered my truck with a flex-fuel engine, I could burn relatively low octane, ethanol-diluted fuel. I averaged between 25–27 mpg. So from a fuel economy standpoint I could afford to take her on extended road trips. In any event, my '58 has served her purpose well. She established my initial credibility, so that during my travels I was able to talk with hundreds of gearheads; they told stories, and I listened. As I compiled these stories, I found a good deal of wisdom among builders and custom owners. They are thoughtful and passionate about what they do, and they're willing to speculate about what it means and what lies ahead.

Gearheads can talk forever about their builds, and the many lessons they learned along the way. Everyone has his own take on the right stance, a good silhouette, and how things should perform. So with that caveat, what follows is a ride based on personal experiences, lessons learned, long hours of interviewing, and some 40,000 miles of road trips across America in search of gearhead wisdom.

3

My '58 Build Story, or How I Was Adopted by a Chevy Apache Behind a South Dakota Barn

...the Platonic Ideal of truckness: that scenario in which I would fix everything, make everything perfect, put it all back together perfectly, and thereafter drive and enjoy my perfect truck. —John Jerome, *Truck*[1]

Although I had no idea at the time, my "build" odyssey started in August 2002 when I paddled my kayak into South Dakota's West Whitlock State Park. There is a much longer story here, but the short version is that I was kayaking the Missouri River, working on a Lewis and Clark related guidebook.[2] Ultimately, I paddled 2,321 miles solo from the headwaters in Montana to St. Louis at 3 mph. Along the way, I saw some really sweet vintage trucks. Some were still in use, but many had been abandoned along the riverbank, nose first in the water, as an informal effort at erosion control. As an admirer of old trucks, there were times that I wished I had a socket set, torch, and a way to carry away a boatload of loot. Of course that was not to be: I was on a mission, and my toothpick of a watercraft was not an appropriate classic truck parts collection platform.

As an old-school geographer—a "field guy"—I focus mostly on landscape change; documenting what's lost, what's gained, what persists, and why. So as I paddled past the old cars and trucks dumped unceremoniously along the Missouri River's banks, the idea for this book began incubation. I decided to re-purpose an old vehicle and drive all over America collecting stories about people who still remain passionate about rebuilding, modifying, and driving old cars and trucks. In essence, my ride would be my calling card for conversations with gearheads and hot rodders.

Fortunately, I have some mechanical skills, because building an old vehicle requires either a lot of discretionary income or some pretty solid wrenching talent. In my earlier lifetime, I pumped gas and worked in a garage, where I learned the basics and, eventually, how to rebuild engines, do bodywork, and how to paint. With this context, what follows is how I came to own a 1958 Chevy Apache Fleetside long bed pickup, which I hot-rodded to use as my vehicle for conversations.

My truck came from behind a red barn near Gettysburg, South Dakota. I was drinking a beer and eating deep-fried chicken gizzards at the local hangout with a group of walleye fishing guides and park staff when we somehow got on the topic of old trucks. I probably mentioned some of the old metal I had seen while kayaking the Missouri River; in any event, we were talking about our favorite trucks. I volunteered that my favorites were the '58 and '59 quad headlight Chevy Apaches. Jerry Gray, one of the guys at the table, said he knew where there was one for sale. After our meal, Jerry asked if I'd be interested in going with him to see a '58 Apache that he had bought for $1,200. And he added, "I really don't need another old truck; if you want her, I'll sell her to you for what I'm paying." Two days later Jerry and I went out to see the '58.

My first impression was that the truck was pretty rough. It was the three-quarter ton model, not the more desirable half-ton. However, as I looked her over more closely, I could see that the old girl was actually pretty solid. She had a nice patina along with the wear you would expect on a well-used farm truck—mostly surface rust, dings, dents, and a few small tears. Her driver's bedside "rocket" was flattened at the back, bed wood was totally gone, windshield shot through, and the remaining windows cracked badly or busted out. A large bullet hole graced the passenger floorboard. The bench seat had hardly any fabric or padding, and chicken manure covered the floor and most of the dash. She was beautiful.

On the way back to my campsite, Jerry said to me, "If you want her, she's yours for

My 1958 Chevy Apache Fleetside "barn find" located at Gettysburg, South Dakota (August 2002).

the $1,200 that I paid. Think about it and let me know in a couple months." A couple days later, I drove back to Upstate New York. Two months after that Jerry called: "So do you want the '58?" I probably hesitated for a few seconds, and then said, "Yes." Never mind that I had zero plan for getting the old girl back East. In retrospect, I have no idea what I was thinking. Maybe it was like what happens to people when they "just visit" a dog pound and come home with a cute puppy. Except that I ended up adopting the equivalent of an arthritic St. Bernard. In any event, I now owned a truck that was nearly a half-century old and halfway across the country. At this point, I believe, I began referring to her as a "classic" truck. This seemed to help to suppress the recurring notion that I was a real idiot.

I shopped around for transport and talked about my options with Jerry. He volunteered that, since he had not seen the East in fall colors, he'd bring out my low-buck trailer queen if I'd pay his expenses on the way out. I agreed, and in October of 2002, Jerry brought the '58 to my home on a U-Haul trailer.

I got her running after I drained and flushed the gas tank, cleaned the fuel lines, and rebuilt the carb. She fired right up, smoked a bit, but settled down to a good idle once I adjusted the carb. Over the next year, my son Daniel and I rewired her, and managed to get all her parts working or replaced with an assist of several classic truck parts suppliers.

My cleaned up, nearly street-legal '58 at my home in Cortland, New York (October 2003).

By the following October, I got the title worked out, bought insurance, and purchased historic plates. I put a muffler put on, had her inspected, and got her on the road. She rattled something fierce, steered like a boat in crosscurrents, but her brakes worked. She was marginally legal. I kept her in her half-century old barnyard patina, with the exception of the couple inches of chicken manure inside the cab. After a power wash of the latter, I pulled the bench seat out, stripped it down to the frame and springs, and had it reupholstered.

So, a year after purchasing her, she was good to go. Except that she was very loose in the front end. She drove like an old tractor. In retrospect, she was probably only safe on back roads followed by another vehicle with flashers on. By the next year, I had my epiphany—my sudden moment of realization when the effort shifted from restoration to custom. She was a three-quarter ton model, which made it difficult to get the parts necessary to really make her roadworthy. Her 4:57 rear end was an issue; even if I'd managed to repair the front end, she'd never be able to cruise at highway speeds.

Of course, there is the issue of "respecting" the old metal to sort through. Once the decision is made to alter a stock vehicle and the first cuts are made, there is one less survivor of that year's production run. This truck had spent most of her life in farm fields, and on backcountry roads carrying sacks of grain and fertilizer or loaded with fence posts and barbed wire. With 68K on her odometer, she was worn down but not totally worn out. She was restorable. With the exception of a rusted out rear cab corner and some lower door rust, she was solid.

As I reflect on my decision to build a custom cruiser, I did struggle with the thought that carving away on a "historic" truck can be viewed as a violation of some sort, an irreverent act. Nevertheless, by this time I was committed to writing an interview-based book about hot rodding, and needed a vehicle for the effort. So I decided to create a new truck under an old skin. My hope was that she and I would develop the karmic understanding necessary to complete the transformation, and that the replaced drive train and suspension parts would yield a capable union of old metal and new technology. She would return to the road transformed, capable of making previously unimaginable distances with grace. She would be able to run all day at the highest posted speed limits and would purr like a kitten down the highways and back roads of America. That was the plan.

Things got serious. I removed the rear axle and suspension and replaced it with a new Currie 9-inch Ford rear axle with a 3.25 gear ratio rear member hung from a TCI 4 link suspension. I researched engine and transmission options. My original intention was to power her with a small block Chevy engine with a four-speed transmission. My plan changed after I read an article in the July 2005 issue of *Classic Trucks Magazine* that discussed putting late-model Chevy LS series engines in old trucks. The article, titled "Practical Power," cited dependability and high performance with low emissions and good fuel economy as advantages.[3] About this time I was also communicating with John Gilbert, editor for *Custom Classic Trucks* magazine. He wrote to me that "things have evolved to the point that the smartest move a person can make these days is to drop new technology into an older truck and enjoy the benefits of both worlds."[4]

So in mid–October 2005, I purchased a Chevy 5.3 liter Vortec engine with a 4L60e transmission from a wrecked '05 Avalanche with 13 miles on it. For less than $4,000, I got a virtually new engine and transmission, complete with an intact wiring harness, computer, electronic throttle by wire gas pedal, and all engine accessories. So the proverbial die was cast: I would build my '58 with ultra-modern technology.

I sent the wiring harness to Speartech Fuel Injection Systems in Anderson, Indiana, for reworking and reprogramming the computer. I pulled the 235 straight six engine and transmission and removed the front suspension and steering linkage and then retired from my project for the winter. The following spring, I installed a Mustang II front suspension kit with disc brakes and power steering. Once everything was together, I put on a set of Torque Thrust II wheels and had a rolling chassis.

I made a list of what I needed to do and tacked it to the wall in the garage. It was a long list; my progress was slow. Early in the effort I came to appreciate what experienced builders already know: Most jobs take two to three times as longer than planned. I often had to order additional parts, and then wait for a week or 10 days for them to arrive. I typically had two or three jobs underway at the same time, which created a sort of start-stop-start, multitasking quality to the project.

My progress at actually crossing off tasks on the list was discouraging. But, early on, I promised myself that, when I finally got her legally on the road, I would head straight for the A&W drive-in. Once there, I'd order a couple mugs of frosty root beer and two chili dogs. So one step at a time, assisted with a seemingly continual stream of parts brought by the UPS truck, I began to see progress. And every time I drove past the A&W, I said to myself that I'll be driving in there with my '58. I was on a mission; every time I crossed off a task, I was visibly closer to the goal. Then the A&W burned down.

My plan for a celebratory meal at the A&W was torched by a kitchen grease fire that got out of control. Everything inside the restaurant burned. For a while, it seemed that the landmark drive-in would be bulldozed. No other drive-in was even remotely close. Somehow, until the fire, this one A&W had survived. My dream to celebrate "mission accomplished" by a cruise into a real drive-in seemed totally trashed. The A&W sat silent, windows boarded up throughout the summer. I wondered if this was some sort of message about the folly of my enterprise. By fall, however, rebuild of the A&W got underway. My mission was resuscitated.

I attached the reworked engine wiring harness and worked through the puzzle of how to mount the engine and transmission in the correct position. It was now four years since I'd entered into partnership with my '58. She was coming along slowly. I still had not fired the new engine, but at least I had the big parts in place. It was time to sweat the small stuff. With the engine and transmission installed, I ordered a set of tight tuck exhaust headers to provide room on the driver's side for the steering column and shift linkage. I made some rough measurements, and in my mind's eye, I could see how this might fit together. I worked on her at night, through the fall and early winter.

When I buttoned the '58 up for the winter of 2006, I was pleased with my progress. Big tasks remained—connecting the steering, rewiring the truck, obtaining a custom-length driveshaft, and getting all new glass—but I was well on the way to finishing my build. I was at the point when the sum of the parts began to resemble a whole truck. Things got more complicated, as any gearhead can appreciate. I needed to find the sweet spot where everything functioned together smoothly. Change one thing, and everything else changed.[5]

Chevy LS engine with 4L60e transmission being lowered into place (August 2006).

As I worked through all this in the spring of 2007, there were times when I had to put my tools down and just walk away. I came to accept that (1) there is no profit in continuing to press against a mounting tide of frustration, and (2) determination is no substitute for finesse. The '58 presented me with many opportunities to practice the "put the tool down, think about it, get a Band-Aid for the knuckle, and come back later" lesson. I have learned that you cannot rush the process because this approach will not result in a well-crafted solution.

Eventually I got everything into position. I added a pair of leather bucket seats from an '03 Silverado truck and completed the wiring with a Painless (that's the company name) wiring harness. By the end of May, I had all the basics wired up, including a set of Auto Meter custom gauges. With everything connected—oil pressure, water temperature, and voltage gauges—I was ready to attempt ignition. I checked all the fluids and hoses. I added five gallons to the gas tank. I turned the ignition key to the first stop and heard the fuel pump hum, indicating it was pressuring up the lines. I checked for leaks and found one at a connection to the fuel pump. I tightened the connector and stopped the leak. Everything looked good. I turned the key back to stop. It was late, and I decided to let things sit overnight.

Early the next morning, I took a deep breath and turned the key to start the engine. The engine turned over once and fired right up; I checked the gauges and saw the oil pressure come up. The engine ran a bit rough, but then settled in to a nice idle. I shut her down and inspected all the lines for leaks. No leaks. Since this engine had been sitting for two years before I was able to start her up, I was relieved.

From this point on, things came together quickly; brake lines, new windshield and back cab window. I installed one-piece power windows in the doors, changed out the taillights to LED versions and added halogen headlights. Finally, I carefully measured for a driveshaft and ordered one from Inland Empire (an Ontario, California, custom driveshaft fabricator) and they quickly shipped one out. I attached it, double-checked everything, and drove it down and then back up the driveway.

It was now August 2007. She needed an exhaust system. After some checking around, I had her towed to Randy Tubbs at Cayuga Performance for an exhaust system.[6] Randy did a beautiful job. In addition, he reinforced the MII crossmember and went over all the welds to ensure that the front suspension was bulletproof. She passed inspection, and I drove her home without any issues. By the end of the month, I got the semi-finished bedsides on her, insulated the floors, and installed carpet. In mid–September, I took her to the nearby Little York Car Show and, afterwards, to the Cortland A&W Drive-In for bacon burger baskets and mugs of cold root beer. Mission accomplished.

Well, "mission accomplished" almost. She was in primer. The air conditioning and the cruise control needed to be hooked up. She had no bed wood, needed more bodywork, and was still in rough primer. Still, her big V8 purred. She ran well, steered easily and true. But she still clanked and rattled. The doors were the main issue; the hinges needed replacement, along with the weather stripping. So loose ends remained, but after four years, my '58 was finally on the road, a true highway cruiser.

Before putting her away for the winter, I took her to my friend Rick Bisutti's shop and we started the process to get her in paint. He worked on the bed over the winter, getting it ready for paint. For color, I settled on what might best be described as mint green, a light color that would hide the old truck's small imperfections and make it easier to repair some of the chips and dings earned by long highway distances.

It had always been clear in my mind that she would be a driver, not a trailer queen. My design was that, regardless of how many miles driven, she'd clean up well enough to be presentable. She wouldn't win a Best of Show trophy, but she would have a solid grace about her, reflecting her ability to go any distance.

In the spring of 2008, Rick painted the bed. It turned out to be much brighter green than I had imagined. Rick called me after he had sprayed her to say, "It's really green." He wanted me to come over to look at her. I think he was a bit nervous that I might not like the result. Under the paint booth's fluorescent lighting, the color looked like baby shit green. However, once we got it into natural lighting, the green color worked. Much to our relief.

Shortly after that, Rick finished prepping the cab. He sprayed it and, once the paint was dry, we bolted the bed on the frame and my initial build was complete. More or less. It would be another two years before I had everything dialed in well enough to allow for my first long-distance road trip.

4

"Drive What You Build" Lessons

But I had a running truck. I floated into the house to regroup around a peanut butter sandwich. Euphoria began setting in. I did it. I figured the son of a bitch out.—John Jerome, *Truck*[1]

It is not about speed. This is the first lesson. Nothing is quick; everything takes longer than you expect. When you are a home-build guy working with old metal, you quickly learn that plans are subject to frequent change. I've seen some of those TV shows where car or truck builds are completed within an insane deadline. There is often some drama. Of course, these builds are accomplished by pro-grade garages with highly skilled workers, all the necessary equipment, unlimited resources, and a lineup of suppliers eager to have their products showcased. With the exception of someone living in Los Angeles or Kansas City—where major suppliers of classic truck parts are concentrated—a classic truck builder can rarely get parts locally. So there are delays; back orders are common and their notice sometimes arrives with a partial shipment. From the standpoint of the home garage custom builder, the TV build shows are anything but reality TV.

Most "home-built" projects have limited budgets and can take years to complete. My project has spanned seven years. I had to save money for major expenses such as rear suspension and axle, front suspension, wheels, and power train. I first began working toward a rolling chassis. In between, I worked on the smaller jobs. So things were done more or less in a logical sequence, as my savings allowed. And my efforts were seasonal, with the project going into hibernation by late October most years. Of course, because I had a real job, I worked on my truck on weekends and late at night whenever I could squeeze in some garage time. This is what gearheads do. If I lived in a warmer climate or my garage was heated, had adequate resources, and worked at it full time, I might have finished the project in a year. Or three.

But a home-built cruiser is a project, a process, and a long-term relationship. You have to love old metal and be passionate about craftsmanship in order to construct a functional highway cruiser. Building a ride is not a simple assembly of parts. A good deal of skill is necessary to engineer a well-performing ride. The effort also requires an eye for design and a penchant for precision. Where do "do-it-yourself" builders get these skills? And why do they pursue this far from "plug and play" activity? What is the source of their passion? How can custom culture persist in the face of ever-rising costs for materials, unpredictable fuel prices, competing distractions, and in light of negative environmental impacts related to pollution?

In my darker moments, I cannot help but think that the future has no place for old men and old metal. Perhaps today's hot rodders represent a passionate commitment to what is going to be gone soon. Maybe the only road ahead is a virtual highway. Perhaps we are just a bunch of old fart, graybeard baby boomers taking one last shot at the past. Are we wallowing in a '60s and '70s muscle car nostalgia financed by retirement savings and resources cut loose by empty nest home downsizing? Or is this a final convulsion of a "most toys wins, last man standing" one-upmanship contest? Are we participating in one last before-the-enviro-crash, super-display of conspicuous consumption?

But as I said, these are thoughts of darker moments. Mostly, I think, home-built cruiser folks are seeking some kind of relief from "affluenza," essentially a high-speed consumptive lifestyle.[2] Maybe home-build guys represent a resistance movement against the pressure to have the latest car, truck, or thing. Perhaps old car guys and collectors are really shrewd investors at a variety of levels.

Members of rodding culture build unique rides. They resurrect and modify old cars and trucks. They learn to be mechanics who can troubleshoot problems. They puzzle out issues, take old things apart, and get them back together so they run well. They are passionate about old metal, and they share their knowledge and enthusiasm with others. There is a social element to the activity. Cruising, attending shows, and helping friends with their cars are shared experiences. According to Harvard social psychologist Daniel Gilbert (aka Professor Happiness), "people tend to take more pleasure in experiences than in things." One reason for this, according to the professor, is that experiences tend to be shared with other people and objects usually are not.[3] Perhaps investing time and resources in building and modifying custom rides represents "wise shopping" for happiness.

So, in my lighter moments I think, just maybe, hot rodders are onto something. They are both potentially a force for social integration and a source of wisdom about how things work. They recycle old metal and used parts to trick out their rides to increase performance. This reduces environmental impacts. At a time when fuel costs are first going up through the roof and then crashing into the basement, and when our high-on-the-hog lifestyle seems coming home to bite us, perhaps home-built guys may have exactly the skills necessary to keep us moving down the road. Happily, in old metal, with style and relative efficiency.

Part Two

The Run Out

5

Introduction to the Interviews

> *It is no hyperbole to suggest that this book is a result of a collective intelligence and curiosity.... In this book are a hundred American voices, captured by hunch, circumstance, and an estimate. There is no pretense at statistical "truth," nor consensus. There is, in a manner of jazz work, an attempt, of theme and improvisation, to recount dreams, lost and found, and a recognition of possibility.*—Studs Terkel, American Dreams, Lost and Found[1]

I began interviewing in an organized way at the 2009 Nashville Goodguys car show. It was June, and I swear the early afternoon temperatures were 104 degrees or more. Gary Meadors, Goodguys founder and chairman (1939–2015), had set me up to shadow Bill Goodwill, who oversaw the awards team. Bill was kind enough to allow me to accompany him in a golf cart as he drove around looking for the nicest vehicles at the show. As we rode around together for two days, we got to know each other. We hit it off and he responded to my questions graciously; I learned a lot from him. In addition to helping me to understand why he was picking certain vehicles, he shared his perspective on how and why guys get involved in the hobby. Bill gave me perhaps the best explanation for the wide range of motivations and attitudes among people bringing their hot rods to shows:

> Why do guys do this? It's their hobby; it's their love; it's their golf game; it's their fishing or hunting. This is what guys do instead of RVing. You do whatever makes you happy, you know? To give you a hobby or sport to do in your spare time. This is their leisure time; this is what they choose to do. The fellowship is a big piece of it. A lot of times guys get into it initially to win awards, to win trophies ... after a while they find out their car wins a few times. Then it's done, people won't pick it..... Then they either build a new one so they start getting more trophies or they start driving it and they find out, "Hey, this is the real fun."
>
> It's amazing how many guys I know who got in to hot rodding to win trophies. It's like competition hunting or fishing or anything else. You know, sometimes you get in it that way and you find out this is real work. And a lot of them got into this because they had friends who went to local cruise nights and after a while when they went, "Jeez, there's a lot of cars that are nicer than mine." And then they start up. It's like getting an RV. You start with a pop up and you end up with a 35-five foot double wide, slide-out model. You see guys start at the top end and go down and you see guys start at the bottom end and go up. And that's kind of how this goes.

As I reflect now (2017) on the many interviews that I have collected, there are certain ones like Bill's that echo the themes of this book. I have included a selection of these.

My "finished" truck at the Goodguys 4th Great American Nationals at Pocono International Raceway (September 2008, Long Pond, Pennsylvania).

Typically, I had a series of questions that I brought to every interview. I began by asking permission to tape the conversation, and frequently sent them a transcript to check for accuracy.

Most interviews began with the question, "So why hot rods" and moved along from there. My follow-up questions typically focused on "What was your path, how did you get here and what's the future ahead?" Beyond these starter questions, the interviews were freeform. Rodders talked; I listened. A lifetime ago, I was a geography graduate student preparing for a year of fieldwork in Mexico's Yucatan. My adviser's last instruction was, "Go down there and see what you can learn about the fishing and prospects for coastal resources. Come back and write it up." That worked for me, resulting in my doctoral dissertation. Throughout my career, I have successfully employed this approach. Maybe it helped that I have a graduate degree in counseling; my friends will tell you that I "connect" with people. Be that as it may, I have always found that if approached respectfully, people are willing to share information about the evolution of their interests, passions, and motivations.

My original idea was to publish a book of interview-based stories about hot rodders. However, the more I listened to rodders and reflected on the significance of their stories, the more I was compelled to try to understand the activity's larger context and consider its prospects for the future. In the final section of this book, "The Run Down," I focus on what I learned from rodders and why it matters.

As you read these stories, you will find a rich cross-section of experiences, motivations, and opinions. Every story communicates incredible passion for the activity. What follows are stories that most helped me understand "how we roll." I worked the interviews into a "story" format, at times rearranging the material for flow and relevance while trying to accurately reflect what was said.

Many stories did not make it into this book. I left out those that were similar to material elsewhere. Whether or not a particular rodder's story appears here, I remain deeply thankful for the time and insights each person generously gave me. Virtually every interview was critical to the evolution of my thinking and material from excluded conversations shows up in other sections of this book. Finally, if you hang around hot rodders for any length of time, you know that many of them are pretty good bullshitters; colorful expressions are typical. That's just part of how we roll. In most instances, I have preserved the flavor of the conversations.

Hot rodders design, build, run, fix, and rebuild their rides. As you read these stories, you will come to appreciate that everyone's gearhead gene expression is unique. Skills, motivations and levels of participation vary, but the constant is total passion for the enterprise. Every one of these stories is about someone who is defying the trend toward mechanical cluelessness, in defiance of the contemporary.

6

Bobby Alloway (Louisville, TN)

Bobby Alloway owns Alloway's Rod Shop in Louisville, Tennessee. He built his first street rod in the mid–1970s. A decade later he built a '33 Victoria that won the Ridler Award. He is, by any yardstick, one of the top builders in the nation. And one of the most gracious people in the business. I first met him at the 2009 Nashville Nationals when I shadowed the Goodguys' awards team. I interviewed Bobby at the Columbus PPG Nationals; we sat on lawn chairs in the back of his car trailer. He is straightforward, has a wry sense of humor, and clearly appreciates the help he has received from other builders. For most of his career, he has built street rods, which he defines as built from a vehicle of the '30s or '40s. His rides are sleek and wicked.

BOBBY ALLOWAY: I started fooling with cars when I was in high school. By the time I graduated from high school, I'd had 22 cars. Now maybe 20 of them may not have run, but I had 22. I fooled with '55, '56, and '57 [Tri-5s] Chevys. I just loved those. My dad worked at a dealership in the late '60s early '70s and back then you could pick up a '55 car for $150 or $200.

We had a big old back yard and had a fencerow full of five, six, and sevens. Well, I ended up trading them for a T-bird. I am sort of ashamed of that now, but I traded. And repainted it, put in an interior and put a fake blower on it. I was a kid then, but it looked mean. A guy into Fords tried to trade me a steel '33 three window for it and I wouldn't trade for my T-bird. Now that's how stupid I was back then.

I graduated from high school in '73. I've always had a love for cars. My dad worked at a Chrysler dealership, and they had sold a dealer financed, '70 AAR Cuda[1] to a guy up here in the mountains, and it was a pretty rough place. Nobody wanted to repossess it, so the owner of the dealership told me if I'd go repossess it, I could have it for what the dealership had in it. So me and my dad went and repossessed the Cuda.

Well, the guy was in a house trailer and when we came up he met us at the door and handed us the keys. I mean it was one of those deals where you thought it was going to be a bunch of problems, but it wasn't. There was no shotgun. So anyway, when I was a junior, I had a '70 AAR Cuda. And then I wanted a Corvette real, real bad. And I sold the '70 Cuda, and ended up buying a '71 LT1 Corvette, which I still own.

When I did that, I couldn't afford to drive it. I was paying for it and it set there and

Bill Goodwill (Goodguys Nashville show director and awards judge), left, talking with Bobby Alloway (June 2009, Nashville, Tennessee).

I still drove a '55 Chevrolet. Anyway, a friend of mine who now works for me—Joe Bailey who has been a friend forever—talked me into building a '32 Ford. I still had the Corvette, but at the time I was fooling with Tri-5s, and he said I need to get in a street rod. I couldn't even understand why they were fooling with street rods, but anyway, the first street rod that I actually built was a '34 Ford. I bought it. It was an old ex-racecar and had number 78 on the door. It was a dirt track car. Yep, number 78, the doors had been welded shut, the deck lid welded shut, and it had nothing in it. No window mechanisms; just stripped out. So I bought an old T-bird and took out the seats and the windows, all of that out. And put it in the '34.

Now, this was back in '74 and '75. Back then there were no kits, none of that. The old timers were into restorations and didn't want you to hot rod those cars. So when you bought a part or a car back then, you just about had to lie about what you were going to do to it. You couldn't be putting it on a street rod because it was just taboo to put a good part on a street rod. And so I did that car and Gray Baskerville, of *Hot Rod* magazine, shot that car. And I was hooked from then on. Gray was a great guy and a friend until the day he died [2002]. Me and my wife, we'd visit with him. He was the one that probably actually got me really hooked on street rods. And shortly after that I was building another '34 Ford. I am trying to say '33s, but they were actually '34s, back then. Everybody wanted a '34 instead of a '33. The grille was bigger in the '34, it had more chrome on it, but I don't know. The '33 grille is so much prettier; the offset is just right, you know, so it's swapped around. Of course we were running tail lights, and bumpers, and all of that stuff back then, door handles, door hinges.

But Gray introduced me to Boyd Coddington, who then worked at Disney. I needed a dash. Nobody made a dash, and I had this idea that I wanted to make an aluminum dash. And I wanted aluminum door handles, and he said this guy could do it. So I called Boyd, and sure enough he made me a dash and he made me a set of door handles. I never met him until 1980 at an Oklahoma show. And from then on we spent quite a bit of time together. I talked to him on the average every day just about, until three days before he died [2008]. He was to the point he couldn't talk. And I talked to him every day he was in the hospital; anyway, we were real good friends.

So I got hooked up with Boyd and—later on—Pete Chapouris [co-founder of Pete & Jake's Hot Rod Parts and until recently President of So-Cal Speed Shop]. And because I am from Tennessee, I would go out there to see what they were doing. I'd get back home with ideas that were two or three years ahead of what the local guys were doing. Well, the first time that we shaved everything off the car—took the door handles and the hinges and all of that stuff off—back in Tennessee they still weren't doing that yet. We were still into the bells and whistles and all that stuff. We took a real nice three window to a show and didn't win a thing. And I couldn't understand it because the car was so good. The car was excellent. After the show I asked, "What in the world was wrong with my car?" And they said, "When you finish that car, it'll be a great car." And I said, "What do you mean when I finish that car?" They said, "Well, the door handles and all that stuff wasn't on it." So then I asked why my car wasn't the best unfinished car there. And I was told that it was "Too finished to be unfinished." Anyway, I sold that car and two years or so later it came into style. I learned a little lesson about that. But still, that was what I wanted to do. And that was back in the early, early '80s.

Then I sort of got burnt out for a while. I don't know why. I don't know. I just tired of fooling with it. You know, you just have these things. I got back into motorcycles; I raced motocross for Kawasaki for a couple years. And then I got back in it. I decided I wanted to build real good, so I built a '34 Victoria. And we showed it at ISCA[2] for a while. And did real good at ISCA. I painted it red. Everything before that was black. I painted it red because the magazines were telling me that they weren't going to shoot a black car. So I went red.

And then the following year I built a '33 Victoria and we took that up to Detroit for the Ridler. I spent every dime that I had on that car. We got up there and I met Tom McMullen from *Street Rodder* magazine. And I thought, "Here I am: A nobody. Nobody knows me, and here he is, hanging out with me for the weekend." I was broke. I remember a hamburger at [The Hotel Crowne Plaza] Pontchartrain was a dollar. And that was real high for us. And, to make a long story short, we won the Ridler in '85.

And really needed to. Me and my wife—she was my girlfriend at the time—we really needed the prize money. We didn't have two nickels to rub together. I mean, gosh, it was $5,000. I just wanted it real, real bad. And I had done that car by myself, done every piece of it with a little help from some friends. Why? I don't know; I just had this show bug. And then we showed that car a little bit after the Ridler, and then I went to an indoor show and got pounded on pretty good and quit. After that I mean. I quit.

6. Bobby Alloway

I'll tell you a funny story. Me and Boyd was friends at that time and he was showing on the West Coast ISCA, and I was showing back here. And he was showing Larry Murray's '33 Phaeton at those deals, and we were winning back here and they were winning out West. And we were going to butt heads at some time; which was fine.

We had won the Ridler. Then we went to other shows chasing points. And we went to Philadelphia, I'll never forget it, and there was an old car there, it was the "Renaissance Delivery" [Ron Barnum's 1983 Ridler Award '29 Ford 3-door Sedan Delivery]. And it beat us. And I never could understand why. I mean, I could see stuff that was wrong with that car. After the show the judges came to me and they said they wanted to explain this, and I said, "If it's up and up, you don't have to explain anything." But they said there were some things wrong with my car and if they continued to let that go, then when we got to the grand finale, it would hurt me. I said, "Well, why didn't you just tell me that instead of giving first place to an old build."

That night I was a little upset, so I called Boyd. I said, "You ain't gonna believe what happened. We got beat at Philadelphia." And his comment was, "Well, you win some and you lose some." And I thought, "Well, you dick, you, for saying that." And I said, "Well, I understand that he's coming to Omaha next weekend after you." And he says, "Well, we'll see." The next weekend I get a phone call on Sunday night and he's cussing, saying, "You can't believe what's happened. The same car beat Murray's." And my comment was, of course, "You win some and you lose some." So anyway, to make a long story short, I quit. I sold my car and made a down payment on the house we're living in now. Then Boyd kept showing and he ended up winning the America's Most Beautiful Roadster Award with Murray's car that year.[3]

So I got out of it for a while. I probably was out for four or five years. I had a real sour taste in my mouth, just about car shows in general. You know how you get like that. You just get burned out. I fooled around with four-wheel drive, for a while. Off-road stuff and jet boats. Got into boats for a while and then decided to do that. And I did. Then back in '88 or '89—I guess 3 or 4 years later—I had a bunch of parts around and I put together a three-window coupe.

Nothing happened in particular; I was just back again. Then me and Rat's Glass[4] hooked up, and we built a fiberglass body three-window coupe. That was back when fiberglass bodies hadn't taken off so good. So I built one. I still have the first one. And that got me back into it pretty heavy.

I was still going out and visiting Boyd. We talked a lot. Boyd once told me that he was either going to get twice as big or twice as small. And he chose to go twice as big. And I chose to stay where I was at. I am not so much into that California lifestyle. I love the cars, but I still have the family, and I don't want to work offshore. I mean my shop's at home. That's where Boyd started. If I was to expand, go to a big storefront, I'd have to charge more. And then you're not charging for the work, you're charging for the building, the atmosphere. You are charging for a look that has nothing to do with the car.

So sometimes I go to these real nice shops and I think, "Man, I'd like to have that." But then I don't want the mortgage for that, because you have to pass that payment along. And the cars are too expensive. Stupid expensive. And it's ... I'd rather try to keep the price down and do the quality car. And we can do it where we are. I mean, we've proved that over the years. So I don't want to get any bigger. I might want to get smaller, but I don't want to get any bigger.

There's still a lot that you need to have. And my guys at work, we don't work overtime, because they've got a family. Once they give me eight hours a day, five days a week, they need to go home. And most of my guys have cars themselves that they work on, and that's fine. That keeps them involved, and it's sort of like a family in our shop, but I don't want to get any bigger.

Did I ever dream that I would be at this level? I don't know, I don't think I am at a level. I got inducted into the Rod and Custom Hall of Fame,[5] and still haven't gone out and got my credentials yet. Darryl Starbird is a great guy, and I apologize for not going out and getting them, but I don't have the time. When I am not at a show or building a car, my daughter plays softball. And we're at the game, or I am doing something with the family so it's real hard to get out there. But when Darryl told me I was to be inducted, the first thing went into my mind was the thought that the Hall of Fame was something real big and real good. And either they have run out of people to put in there, or it's not as big as I thought it was because they have inducted me into it. And as Boyd said, I am "out in the sticks whittling stuff out of Bondo and duct tape." So I am very appreciative.

Author David Miller with Bill Goodwill, awards judge at 4th Great American Goodguys Nationals car show at Pocono Raceway (September 2008). The event was nearly rained out, so we made plans to repeat my "shadowing" effort at the 2009 Nashville Goodguys show.

6. Bobby Alloway

Bill Goodwill taking notes about Bobby Alloway's black '33 Speedstar roadster in his trademark "Ohio Flames" at Goodguys Nashville show (June 2009).

But I sort of feel if these icons are my peers, then maybe that drops them down a step, because I know who I am. And I thought they were these great people. And they are in my eyes. But when they put me up there with them it may pull some of them down, if you know what I mean. They must really like my stuff, or they have a higher opinion of me than I do, or I've fooled them. Or it may be a little of all three.

What's the future of hot rodding and custom cars and trucks, and where do we move to in terms of the culture today? I don't know any answer other than I don't think street rods will ever go away. But we're going to see more '50s and '60s vehicles. And with ride and handling components produced by Kyle Tucker at Detroit Speed, the autocross testing, and air ride suspension systems? That stuff is as hot as it gets. But what's going to come after that? I don't know. There's going to be something, because some new innovation seems to come up every three, or five years; who knows what it is?

Man, I don't know what the kids are wanting today. Of course I have one, but we started real, real late in life. I'll be 54 years old this year. And I have a 9 year old. So who knows what she'll want to do, you know? What I try to do, I see some young kid in there, I'm going to go up and talk to him. And if he wants to know anything he needs to come and ask me. I'll tell him anything. I did not have that luxury when I was growing up. I had to fight through it and figure it out. Nobody would tell me anything. And I think it's going more to the muscle cars, you know? The Tri-5 stuff is some of these kids' '32 and '33 Fords. And even so, now the Camaro is these kid's '32 and '33 Fords. The only way we have survived is to try to adjust to what's going on.

And I am fine with it. I still love street rods, but I just built my wife a '63 Corvette and I love it, you know. And I am driving a '49 Ford convertible. We are doing Camaros in the shop. We're doing Corvettes. We are doing more '50s and '60s stuff than we are '30s stuff. But that being said, I think the whole world revolves in a circle, and that the '30s rods will never go away. They may be dead right now, but they died in the '80s a little bit, back when the muscle cars picked up and the Corvettes, the Thunderbolts, and the Hemi cars were strong-strong-strong. They sort of petered out at that time but they came back. Who is going to bring something out that really sparks it up? It's maybe us, and I hope that it is. Or it may be Troy, or it may be Chip, or it may be Jessie Greens. We have a lot of these young kids that's coming out that, jeez, it's hard for us old guys to keep up with them. You look and say, "How in this world did they think of that?"

And that was being said in '85 when I won the Ridler, I was one of those guys. But I am still pedaling as hard as I can and I am just barely keeping them in sight. You have Alan Johnson; I mean, those guys are just great at what they do. And as the saying goes, I am so jealous of them it's awful. But they are all my friends. But that's a good thing, because if they weren't doing this good work, you wouldn't copy some of their stuff. I've always said the highest compliment that you can get is to have somebody copy you. Because if they copy you, they have to like it. They might not tell you that they do, but if you sing good, they like it.

I have learned a lot of lessons. And I am trying to teach some of them to my daughter. And it's real hard, because it took me—and I still don't know everything—it took me fifty-something years to get it in my head what I am trying to get her to do at nine, even with her playing ball. You do everything that you can do absolutely the best. Absolutely the best. If it's not the best that you can do, don't do it. If you are not in that frame of mind, then you wait; you take a day off or whatever, but you make whatever you do the best. And be very respectful to all of these other people and you'll be fine. Just be honest about it.

It takes you a lifetime to build a reputation; it'll take you about 30 minutes to kill it. And in this industry, everybody knows everybody. So sometimes you better watch, you better think before you speak. But if you just tell the truth, you don't have to remember what you said.

7

Bo Huff (Sunnyside, Utah)

In July 2010, I interviewed Bo Huff at his museum in Sunnyside, Utah. I had written him from New York to set up the interview. I don't know what he was expecting; but right away he asked me, "Do you want a beer? And how come you don't have a Yankee accent?" So we hit it off immediately. Bo was remarkably introspective and candid. It was the weekend of his Rockabilly Car Show and we pretty much spent a day together. He was an absolutely gracious host. Although I did not know it at the time, Bo had just received a terminal cancer diagnosis. Bo passed away on July 3, 2015.[1] It is not hyperbole to state that he was a giant of Kustom Kulture and Rockabilly Hot Rod Lifestyle.

BO HUFF: My cars are works of art on wheels. Back in the early '90s a guy interviewing me for a book asked me where I thought custom cars were going. I told him it's like Mickey Mantle, one of the best baseball players ever. In his best year he was paid $10,000 and now a guy that just sits on the bench might make $30 million. I says, I look at the cars in the same way. We've been pushed around; we used to sell our work for nothing. I worked 10 years on one hot rod, not because I wanted to sell it, but when I got ready to sell it, I sold it for $2,500. And that's what the prices were. If you look at the books in the back where they used to sell cars, they would have like a radical, radical, radical, custom for 1,400 bucks. And I actually give credit to Boyd Coddington for bringing the prices up.

What I believe is that these cars are no different from art that they pay millions of dollars for. But this is a rolling art. This is the first time that art is a sculpture that also rolls. You have everything, every type of art combined in a custom car. And you turn the key and you go down the street and get thumbs up.

You know, if I was on a desert island and I knew I would never see one more person as long as I lived, and I had the things there to do it, I would build a car. Just because it's really cool to create things. I've got several cars in my mind that I've never done. I don't know that I'll live long enough, but the ideas that I've thought of, I would really, really like to get those done.

I love what I am doing here. Looking back—and I am not bragging—but if you put all the cars that I have built against any customizer out there, I'd hold up pretty well. I would say Barris brothers might have done more. I've done a lot of cars, but I've never

Bo Huff and author holding a 2010 Rockabilly Car Show Poster (July 2010, Sunnyside, Utah).

got credit for nothing until 15 or 20 years ago. I guess, hell, I've been in it for so long they could not overlook me anymore. If I would have stayed in Southern California it would have happened a lot sooner. But that isn't what I wanted to do. I don't know. I guess I don't like waiting in line. I like cars, but I don't like traffic.

I live and work in East Carbon [Dragerton], Utah. I wouldn't say I was comfortable with the community. I stay to myself. I have my own circle of friends here, plus there's a lot of people that come in from out of town. I get visitors from Russia. People from all over the world have been in this museum. But as far as the local people, there aren't even half of them know what I do. I just go to work or over here or to my house.

Yes, if I were in LA or somewhere California, I'd probably have more people barging in on me. See, when I was in Cypress, California, I had a custom shop back in the '60s. And I had to have a rope up back then, that was back when I was shooting metal flake and pearls, panel paint jobs with flames all that, cobwebbing, and ground glass; that was that era. And actually it's come back to that era. I am doing a lot of metal flaking here.

So are there some kids that have learned to shoot metal flake well today? I think so. But they came up with a different metal flake now than the old stuff. I always used the MetalFlake brand, and for my candy colors I use the brand also. But now they have metal flake that lays down a lot quicker. It does everything that the old paint did, but I've still got to add a little bit of ground glass. They have tried everything, but they have never mastered getting that ground glass in their paint. When you put that on something, it's gonna be so good you could kiss it. That was my trick: to throw ground glass in with the MetalFlake. And no matter how good a metal flake reflects light, with ground glass it is way better. A lot better.

Anyway, I left California. I came here and worked in the coal mines six months. Then I went to Arkansas. I lived down there and I was not building cars. I went to college for a year, and my folks had a house that no one was living in, like an old plantation home. So I went and painted that all up and hauled everything off and fixed it up really nice until my folks decided to live there and it was all ready for them. When I was down there I built my black truck. I had looked for one for about three months. Turned out that truck was actually about half a block from my house. They was feeding cows out of the back of it. I told some locals that I'd built customs out on the West Coast. They said, "Oh yeah." So I finally said, "Ok. I'll just build that truck." So I bought the truck and that's the first thing I built.

Now this is a long story. Well, when I got the truck built, my dad kept saying, "Let's paint that dang thing." And I said, "I'll paint it when it's ready." Because I wanted to paint it black, in black lacquer. I called up Stan Betz of Southern California, who I used to get my paint from. I says to Stan, "I want the blackest paint you've got." He says, "All the black's the same except there's a gold toner. If you can use this certain gold toner, put it into the black, it'll make your black blacker."

So we put gold toner in it. We did that to the roadster here in the museum. And that was done in the late '80s. That paint job looks so deep you could fall into it. And like I say, that's old lacquer. So, I put that paint on it; finally, I painted it. And I painted day after day after day until I rubbed it out. I went to the Little Rock show—I drove it in the rain all the way to Little Rock, a little over 100 miles—and I went in and they says, "Well, you didn't register, so you can't come in." Because I was real bad about that kind of stuff. I never register. I just go with the car; it gets me in. So this was in the morning, early. All day it rained outside and all these guys were bringing their cars in on their trailers and all that. And finally about 10:30 or 11 that night when they's going to shut it off, I walked up to the guy and I says, "Listen, man, my car is sitting out there. Either I gotta go or you gotta let me in." Well, the guy says, "Well, where is it? Let me see it." Then he says, "Oh shit, bring that car in. Bring it in."

I had approximately 15 minutes to wipe it off; no display, no nothing. Next morning, I got up early, and I saw these beautiful yellow flowers, so I went over and nailed this lady's flowers. Nobody home, I don't know if she was or not, I just grabbed them flowers. And I put like two or three white vases with those flowers by the truck. And at the time I had a candy gold rim which was the same color yellow, with a spinner hubcap. And shit, I won best design, best paint, best upholstery, first in my class. There was a guy there who was the head of Jim Ray Chevrolet, Fayetteville, Arkansas. He called me up a week later and said, "Would you like to come to work for me?" So I says, "All right."

So I accepted the job, I moved up there, and I worked like three years for him. And then all at once it changed from this Jim Ray to this new owner who was a power monger man who said, "We are cutting you guys' wages back, because we were going fifty-fifty commission." I said, "You cut mine and I am out of here. And I've been here three years and never had a comeback. I make more money than your best salesman. I make more money than anybody in this thing: So every dollar I make, you make a dollar. So you go

Bo Huff's '29 Ford Model A roadster at his Sunnyside, Utah museum (July 2010).

ahead and cut me back, and I'll leave. And you'll have comebacks, because all these painters say they are good but they can't do it." I was good, man, as a painter.

So what happens, man, is I had this shop kind of lined up anyway. I just moved in and started building cars down there. I did a couple cars and then this guy says, "I'll give you five grand if you will paint my Mustang." That was back in '80; that was a lot of money. That worked out. I started in doing restorations. But one day I say to myself, "I ain't never going to do anymore of this kind of work; I don't want to restore cars, I don't want to fix dents." I went and got on the scale and saw that if I didn't lose no weight, I was gonna have issues.

And then I went through a divorce and got the hell out of there. And I came here to pick up and I been here since then. And all I build here is my kind of cars. And I am not getting rich by no means; I live one day at a time and I want to do what I want to do for the rest of my life. So I don't do restorations. Or, you know, if some guy come in here and says, "I want you to build this muscle car for me." If it was money that I could not stand, maybe I would these days, just for kicks of doing it. But I've turned down so many Pro-Street cars and stuff like that; I don't know, that stuff I just don't do.

I am at a place where I can decide what I want to do. And builders do find it awful hard to turn down work with real money behind it. I've done it, a lot of them. And I can do 'em. I did a car for the head of Rolls-Royce. I did his '57 Chevy, and it was Pro Street.

But by the time I was done, I got it lowered down pretty good, and I framed it and he won everything—a DuPont award, the Watson Trail Award, Best Car of Arizona, all this stuff. But that was the last blackwall car I've done.

Where did I learn all this stuff? Actually, I learned a lot from Stan Robles and those little car magazines. The '50s car magazines. I spent the night with Freddy Zubal in a trailer one time and his brother Gary, who had about nine of those little custom car magazines. And soon as I opened one up, man I just went "Ohhhh." And I was only like in the sixth or seventh grade or something, and I just fell in love with cars and the whole thing. After that, every time I went to the drugstore, I got a new magazine. Man, I was buying 'em. I just loved that stuff. I still love it. I'd always drawn some, but I could see it in my head. It's like that with my kid; he's an artist too. I always consider myself pretty good with flames. But these days, my kid, I just tell him what, where I want 'em, and how I want him to do it. And then I just turn him loose.

A lot has changed since back in the day. Stan Robles, who used to work for Barris Brothers, shows up around here a lot. Well, he got out of it 30 years ago. He tried to get back into it here a while back, and he says, "God, it's too much damn work." Things changed. I use a wire-feed welder now; I used to braze everything. When I first started I was using lead; now I never use lead. Brad Masterson and Bill Hines say lead is the best. Well it is; it's good. I tell them that I am glad that they are keeping it alive, but at the same time there's so many chemicals in lead and I've seen a lot more paint screw up over lead than Bondo or other body filler. But I gotta say that early plastic fillers were junk.

I want to say that there might be somebody in Earl Scheib or Maaco or one of those places that's painting more cars than me if he's been there for 30 years, painting five cars a day. But I would say that in my lifetime I've painted as many or more cars as any man alive. But if you took me right now and put me in California with that water-based paint—all that kind of stuff—and the way they are doing things, I would pretty much have to relearn. I use old-school guns; I never used gravity feed in my life.

The new stuff is good. I am sure it is. But, you know, I learned everything from the age of 14 to the age of 67; why am I gonna change now? One time I did a '60 Plymouth, one of only 11 in the world, a station wagon two-door that I did with flames. I used base coat, clear coat paint. I've done a few that won every award. I can do it, but it almost killed my lungs. I mean, that new stuff is a lot harder on you. So I started getting pissed off at it then, and I used to notice that it'd bubble, so I stocked up the basement here with lacquer, and I was buying lacquer systems, till I can't buy one more.

But now I am doing stuff in acrylic enamel too. You can polish it and it'll hold up better. You can buff it down and polish it. Like I did with the car that I just won the Grand National Roadster Show with—the '57 Ranchero. It's done in acrylic enamel except for the panel on the side and the panel on the top where I've got the cobwebbing and the flames.[2] That's done in lacquer and the scallops. But the rest of it's done with acrylic enamel sanded down with 2000 and buffed. Which is just as good as lacquer and it's not going to be as brittle later on. If you take care of lacquer, like I've taken care of the paint on the '32 roadster in my museum, it holds up.

The lacquer on my truck was the best black you've ever seen, there's never been a better black paint job. But they gave me a ticket when I first got here for being too low. You couldn't slide your foot under it. And they had a "lower law" here in Utah. So I parked behind this house that I owned and I just left it there for three years. And snow was just sitting on it in the winter and just shattered the paint. If I would of took care of it, it would still be just like that the roadster.

Anyway, about the "lower law." So I met this Utah state senator, and I went to the legislature and we changed the motor law in the state. So now we can lower cars. And the following year I went to the legislature about no fenders. And now we can run them without fenders. I just got so many tickets for no fenders. I refused to raise my car up; I just put spacers in it, raised it up, got the ticket signed off, then kicked the spacers back off and took off again. There is some outlaw in all hot rodders. I think I've always been that way. I've always been the guy who roots for the Indians. When the cowboys was killing the Indians in the movies, I was always the guy hoping the Indians would get some of them motherfuckers. It was that authority bullshit.

Back to cars. What am I working on now? What am I building? Well, at the shop right now, go up there there's probably 14 or 18 cars. I work a little bit on them all the time. But the main two I am doing are a '49 and a '50 Lincoln. We call them the "Match Elites" in the magazine. In *Ol' Skool Rodz*, they had one article and Gino Dipol called me yesterday and said that they just sent it to press yesterday. The second article is coming out. And I am supposed to have them done by the Grand National Roadster Show, to debut them there.

I got a lot of cars going. The roadster in the museum took me five years to build. That black one. What I do is … I might chop a top and get that going, and then I gotta make a living over here. And then I got this hot rod right here and then so I do something to it. And then one day after maybe one year or five years, sometimes I do them all at the same time. Then I'll finish. I jump on that one car and just finish it. But I do it. Lot of guys say, "Why don't you work on just one car?" But that's not how my mind works. My mind works in a lot of ways; I can see a lot of different visions at the same time.

I have a vision for a car and I decide to chop it. And I've chopped a lot of cars. I'm at around 39 cars. I am counting mostly Mercs and Lincolns. But I have done a '54 Chevy ambulance—which I thought would look stupid—but it turned out bitchin'. I don't know; I've done a lot. I'd really have to stop and think. But what I do is I vision a car. You check every one of my chops—every one—like the Merc in the museum, this one I had to widen the pillar to lean it back much. Every car that I do is different. I've seen customizers, I am not going to mention their names, but they do the exact same chop, day after day. They know how to do that one chop. But I don't look at it that way. Every car has its own personality. And that's how I do 'em. And I am not trying to brag; that's just the way God give me. That's the way I see things.

What do I see before I start a chop? I see the finished car. My brother-in-law, who has since died, I used to tell him, "I am gonna do this, and I am gonna do that … and so on." I could hear later from him when the car is finished, "This came out exactly how you said." But I can see it in my mind finished. I can see me driving in it. Finished. It is … it's like a picture that comes in my head. And it's not just a picture, it's a whole overall thing and it stays there the whole time. And then while I am building a car, sometimes it changes, you know, and I challenge my ideas. Like that Merc's taillights. The hardest thing I've ever did on a car is those taillights. To make them all even because the

Back end of Bo Huff's custom Merc showing the taillights that were, as he said, "The hardest thing I've ever did on a car" (July 2010, Sunnyside, Utah).

compound curves this way, this way, and this way. Everything is different. I ended up making this little template with a little handle on it, until they all fit. That was a son of a bitch.

You know if you change one thing, everything else changes. That's in a lot of this customizing. I've changed a car by a half an inch on the leaf of a quarter panel, because it had to be done. A guy that worked here one time said, "What the fuck's the difference, an inch higher or an inch lower?" I said it's a lot of difference. If you are taking a car with fender skirts, and it's a little bit higher in the back, don't put skirts on it. If it's got fender skirts, it needs to tilt right. Some cars need to be a little lower in the front; snaky looking. Some cars need to tail drag; some cars need to be pretty much flat. But all of them need to be lowered.

Now some cars don't need to be chopped. That's true. The deal on chopping is, like that little Ford right there [see the photograph of Huff's Custom '57 Ford], it would look fine chopped, but I always liked the tops on those. I did a '42 Ford for the Rockabilly Hound Dog.[3] It's got a grill like something like a Merc, you know what I mean. For three months I looked at it; then I did not chop it. I mean I love those chopped oval coupes but, man, that top was just saying, "Don't cut me up, man." So I did a lot of other shit to it. A lot of times you chop a car and you get a really good look with maybe a chop of 1.5", or 1¾", or maybe 2". But if you chop those same cars the regular 3" it just throws it off. A good chop looks right. Barris' builds are absolutely gorgeous, man. Look at his '40

Bo Huff's Custom '57 Ford. He's always liked the tops on these and says it's a car that does not need to be chopped (July 2010, Sunnyside, Utah).

Mercury coupe, the Nick Matranga Merc. When you chop 'em, you have to section the trunk and all that so it flows away but it all blends together.

 What do I think about rat rods? Well, I saw this one car from the Asshole's Garage; it's got like a Straight Twelve Packard or something in it. The guy's wheels are so bitchin', it's the most…. Well, I offered him a job. Oh man, he's got rivets all around the car. It is so bitchin'. What has it got? Blower on it, or something, and a motor from there to over there. It is a rat rod; they built the frame and everything. It sits low on the ground. Rat rods are cool. The morning after last year's show I woke up hung over like a son a bitch. I had partied all night. I stepped out there and the way they'd parked I just busted out laughing. The son a bitches was parked at random. It looked more than a junk pile; it looked like an aircraft carrier. It was bitchin', man. You'll see that at my show; these are good cars.

 So where do I think things are going, what's the future look like? Well, I can't remember his name, but the guy who does *Overhaulin'* … Chris Jacobs. He interviewed me at the Grand National Roadster Show.[4] He says it was all going new, but now it's going back to old. Well, if it kept going new, I would just keep building my stuff. Because those new cars don't interest me worth a shit. The cars I like is the cars with feelings and guts to 'em.

 My first car was a '27 Model T. And I bought it from this guy, a kid who had started it and then went on a Mormon mission. The body hadn't been worked on and it was a death trap. And I didn't care; I just wanted to cruise it. I got everything going and running and everything. But shit ... parts would fall off it, the wiring I bet would burn up half a dozen times, and I pushed it so much that my leg kinda got run over, you know? Cause I was always pushing it. But that was my first hot rod. I started at that level, but when it was done, it won a lot of big shows in California.

Where do I think it is all going now? I hope that this thing is going in the direction that it is going right now. And that's back to its roots. That's what I hope. Now who knows? You know, however how much I love something, there's always somebody out there that hates it. So who knows the future? But I do think there's gonna be a large group of guys going back to the roots. I have since the rat rod thing started. I have a lot more guys on my side than I had before it started. So back in the '80s or '70s, when lot of these old guys were wearing Bermuda shorts and cowboy hats with what look like a fricking bird hit the front of it, I went back to wearing Levi's and all that. Remember those days? Those old guys didn't even look like they are driving their hot rods. If I am going to drive a hot rod, it just don't look right looking like a clown. So it's all together: the music, the clothing, the cars, the thinking. You can't just be a weekend rodder. Like those motorcycle guys are only weekend guys. Well, whatever.... I guess I am fortunate; I can be this way every day. Whether you like me or not.

What started this to going back to the roots, I'd say, is the rat rod guys. And the rat rod guys, instead of buying everything they could buy from magazines—from *Street Rodder*—they went out and started fabricating a lot of their stuff. And then they go to swap meets and find old stuff. And a lot of 'em run into the old guys who restored their cars, who wouldn't sell you an old body if they thought you was going to make a hot rod out of it. And you'd have to say, "Oh no, I am gonna restore it."

Well, those rat rod guys are going to a swap meet to find the right manifold, or something that's 50 years old, rather than buying a brand new one or something like that. Now, I am not quite like that. I don't mind changing all the suspension and the motor and all that, to where it will get me there and get me back. But I am more focused on the design of the car. I'd like a car to get me home, but I'd like a car that looks good. The aesthetics of the car is what I am interested in, more than anything else. Some guys look at my Mercury, and the only thing that they can think of is that motor. But the only thing I can think of is, overall, how beautiful it is. But that's two different kinds of heads.

I am not mechanically inclined. But a man who is mechanically inclined don't know how to do what I do. I used to build my motors, do all that shit, mash my knuckles, but it wasn't me you know? It was just something that had to be done. I was riding around with a buddy, in probably 1959, and he told me, "You'll never have a hot rod because you are not mechanically inclined." And I remember I didn't say a fricking word, but I thought about it. But I went ahead, and at first, I tried to buck what he said. I figured out how to put the clutch in, figured out how to fix the rear end, figured out all this crap. But I could pull brakes apart and put them on backwards every time. That stuff ain't my shit. But then I could pull an old black Ford that was sitting in an alkali flat down here and I chop it 16½ inches. I could see the car in my mind while I was building it. And it come out wicked.

Right now I've got five people working for me. This winter I had, I think, seven.

Should I stay small because I'll be satisfied with what comes out of my shop? I don't know. I think if I'd had a dozen—if I had the money to pay 'em—if I had a dozen craftsmen that would listen, not just the craftsmen, someone who had a vision, and loved the cars, what they was doing, I could control them just fine. But you get mouthy bastards, or guys that don't do what you say, it just slows you down. If they do what you say—I mean, I got over 50 years in this and I pretty much know what you do—it's good. And so if I tell somebody, "Do it this way" and I have to go somewhere—which I do a lot—then I come back and it's not done right, then all it does is that you got to cut all that shit back off because it ain't right.

Do I think that, right now, I could hire myself as I was 30 years or so ago and it would work out? That's a cool-assed question. I was pretty much uncontrollable. If it ain't my way I can't play. Do I understand there are people like that—really gifted and talented—but they just couldn't work for somebody? I never thought of it that way. The best boss I ever had was the guy at Jim Ray. He pretty much would leave me alone. I think I told him to just tell me what you want done, and let me do it my way. And by the end of the day it would be done. And any time I would go in his office, I don't care if he had 19 people in his office, he would excuse himself from everybody and ask me, "What do you need?" Because he knew I'm the guy who made that shop run. And I didn't abuse that; I'd only do it when I'd need to ask a question.

When me and Rolls worked together, we wouldn't talk. It was like we would be on automatic. He would take off doing something on it—say it was his car—and I would just look down and see, well that needs to be done, so I would just do it. And we would work sometimes three days in a row. Days and nights. Be putting a car together, the seats together, going to the show, park the car, go home and sleep all through the show and just get up in time to pick the car up.

Could I work with a younger version of myself? I like that question. If there was a guy like me? I don't know. Who I can work with depends on the other guy. If it was a guy like Stan Robles, yes. Or like Stella who works for me. He's a real good body man, a real good painter, a real good pinstriper and all that; but he always listens to my ideas. I worked with him 30 years.

I've got patience for people who listen, but I don't have patience with people who don't do what I tell 'em. A lot of my employees seen me in the magazines—these Rockabilly guys all the way from the East Coast. From New York, from Carolina, from Virginia, from Oregon, or from Oklahoma. They come out here because they want to work with Bo Huff on the old cars. And they say, "You'll have to run me out of that shop." Shit. This kind of work is so hard that within a day, soon as it turns quittin' time, man they are bumping their shoulders trying to get out with everyone else. And they don't work out. It's not for everybody. And those school guys, the ones who go to WyoTech? I keep telling them, "People don't spend that kind of money. Send them to a regular shop to learn what they need to know."

What's the future of hot rodding? I am training some people. But we just don't know how much time we have. What's going to happen? I kinda think we are going to run out of cars. Old metal, you know? They're still out there. But the kind of cars I build from, most guys would never even touch. You know those guys like D'Agostino, he buys a pristine car and then fixes it up.[5] I get a lot of these cars out of the ditches around here.

Everything has to be done. Everything. I build on some real mad shit. I'll rebuild doors that was rusted away halfway up.

The future of custom cars is in God's hands, I guess. Because who knows? I have—besides the cars that's at the shop—probably got about two or three hundred cars out the back. But there's a few cars I'd still really like to build. But once I build the cars, I don't really care to drive them all that much. I just like to be able to show them and show the young guys what could be done. The cars that I build, well, I am actually building them for me. But it's also for the guys who appreciate this stuff. Guys who say, "Wow, I like that." If I get that out of the guys who know what they are seeing, I really appreciate that.

Did I ever consider working on any of the newer cars? They don't interest me at all. Although sometimes, when I am driving on the freeway and that, I wonder what a new car would look like with the taillights off and the license plate down in the chrome bumper. I think on that kind of stuff, but there's still some old ones that need to be done. And another thing, if you get a hundred chopped Mercurys together—two from every state—and put 'em at random, you'll see nothing but a line of custom copycats.

Is it flattering that people are looking at my stuff and trying to copy my cars? Well, it is and that's OK. But at the same time, I tell people to use your own head and do stuff. Some guy wrote in an article that you shouldn't even have any more Mercurys because there's nothing else that can be done to one. Bullshit! You bring me a Mercury right now and I'll build you a car that's never been seen. And I won't even think of it until I see it, and it'll tell me. That guy is just a guy with a very little imagination. I don't remember who it was in *Street Rodder*, but a guy one time says there's no reason to even try to build one, there's nothing else to be done to a Mercury. I call bullshit. And there's a lot of politics in this whole car thing, there always has been. I don't go for that shit, you know? You should get what you earn. Not because you know somebody. That's a bunch of shit, you know? That's what I think.

What shows do I go to now? The Grand National Roadster Show, because when it was Oakland, I used to go there. I go to SEMA (Specialty Equipment Market Association) Show a lot, which is not really for the cars that much. Viva Las Vegas, and the shows like Route 66 that I was throwing. Although I don't guess that I am going to be doing that. The one I just threw in Vegas, that's the Rockabilly show,[6] the outdoor Rockabilly show. But for indoor shows, the Grand National Roadster Show is my favorite. Now I usually try to make a car there every year, in January, because that pretty much sets the trend for that year.

Who else is building good stuff right now? Well, there's Gary Chopit.[7] He's a pretty good friend of mine. He was from Long Island, lived in Florida. But he's moving to the West Coast. He builds a good car. There's some young kids building some real interesting rat rods. That's the future right there. Lot of the old guys say, "Rat rods? They are rat traps!"

Rat traps? Hey, I've had a few here every year. And at first, holy shit man, they were so raggedy they had parts falling off. I remember one time I heard this noise—I was up in the second floor—and I heard this noise and I thought dude, what the hell is coming up there? It was a rat rod; it had two 16" snow tires on the front. Shit like that! But then you've got the rat rod guys who try to overdo it. You don't have to add a bunch of extra arrowheads and other crap. But it's good that they do it.

I got a little black '32 here that is nothing more than a rat rod but it's painted. But I would consider my style of rat rod a traditional hot rod built in the rat rod style. Every idea on that car is hand built, the taillights, the bar in the middle. The chop is radical, exhaust pipes unique, the steering ... all that's hand done, everything on that car is hand built. But I painted it. And that car, after all that work, the doors were so rusted out, just real weak, jaggedy metal when I first painted it, when I took it to the Grand Nationals I won first place in the Suede Palace. Plus the car club award. And some guy said, "Well that car's plastic." I thought, "You fucking idiot." But it looked like it was plastic. So that was kinda cool. To start with garbage and get something people would pay to go see: That's kinda cool, I think.

You asked me if I could put a message in a bottle for future hot rodders that would be uncorked 30 or 50 years from now. Holy shit. OK. In 10 years.... In 10 years, I think there'll still be custom cars, OK? The message is, after I am dead, about cars .... I would just hope everybody had as much fun with 'em as I've had. But to use your own mind if you are customizing a car. Use your own mind. You can look at other people's ideas but don't copy other people. Play off their ideas, but don't just flat-ass copy 'em, unless you are cloning something.

You know, maybe I guess I got a negative outlook on life. But I just don't know. The way the laws are, the way they are going to solar type cars, and stuff like that. I kinda think our era is pretty much over. I think we're like, one day everybody was riding buckboards and horses, and the next day everybody was riding cars. Can you imagine the culture shock that was for the old wild west guys? And then one day there's all these cars on the road. I mean, that was a big change and I think we're headed for that. I think we are headed for a change where we're not gonna accept gas cars. Ninety-seven percent of the industry in the world is based on fossil fuel right now. I heard that the other day.

I might have made a lot of money in this game, but I am into design. Maybe what they should do is hire me at the Ford plant, at the Chevy plant—one of those deals—because somebody is designing cars all fucked up. Excuse my language. Especially in like the '80s, the '90s, I mean you could not tell one car from another. They kinda tried to put a little bit of design back in with the PT Cruiser and the Chrysler wagon. But really, man, can you spot a car and say, like back in the day, "Hey man, there goes a '57 Buick?"

With modern cars I give up, I already give up. In the '50s, I didn't know anybody that didn't go to the dealership as soon as the new cars came out and just drool over the designs. And then the early '60s, the mid-'60s; it just started going a little bit worser and worser and worser. Until it got to that square period which was like unreal. It's like, fuck, there's no designs.

A good compliment I had one time, this cat, I didn't know he was from Pebble Beach, he got a flat tire with a '40 Mercury convertible down in Green River. He needed a 16" white wall. He called me up and said, "You got a 16-inch white wall?" And I told him I did. Well, he came up here to get it—to get home on—and I showed him all the cars in the museum. He just stood and looked at my Merc for over an hour. When he left, he tells me, "If I was you, I wouldn't worry about being a car builder; your forte is a car designer." And then he told me what he is; he's the guy who evaluates all the cars at Pebble Beach. Which was kinda cool.

I've got a Mercury up at the shop up there that is esthetically fucking beautiful. It is the most radical custom, and I just changed it again. And I've been working on it off and on. I've had nine friends die while I been building this car. And it's kinda madness. I had a convertible in here—Mitch owns it now, the one with the hood up with the skulls—it won the Barris Award at Santa Maria. And it's won a few different awards. I says, to the Pebble Beach guy, "What do you think about that dash?" Because this dash, man, I painted it from candy red, candy tangerine, candy gold, I put a lime pinstripe in the candy root beer in the top then sprinkled ground glass and sprayed it all, and it's like bitchin', man. And he goes, "The work is very nice and it's very pretty, but I am not into that."

Which is a fair and honest statement. Which is cool, I got no problem with that. That's why there is menus; everybody likes different things. So, when he seen my Merc, he about shit his pants. To him that was probably just a lowrider. But I don't build lowriders. I have respect for those guys. They put a lot more money and time into their cars than I do. But I build custom cars, full-ass '50s custom cars.

8

George Ross
(Thompson's Station, TN)

I met and interviewed at George Ross at the Goodguys Nashville Nationals Car Show in June 2009. I was shadowing Bill Goodwill, one of the show judges and—by chance—struck up a conversation with George, his nephew Carl Risch, and Jack DeRancy. Later in the day we all sat down; this interview is the result. For whatever reason, we hit it off immediately. It turned out that George was a builder of the highest of the high-end builds and restorations. And that he is a great story teller. The result was one of the most informative and entertaining interviews of my project. George is an absolute craftsman: A gearhead's gearhead.

GEORGE ROSS: Nowadays, people look at cars as a conveyance. I can sit in a parking lot and I can't tell you, of the five cars sitting in front of me, what they are. If they are not badged, you can't tell, they all look like Hyundai unisex, they all look the same. There are very little differences. Growing up as a teenager, at night in my '56 Chevy convertible, I'd look in my rear view mirror and tell from the glare of the headlights what the heck the car was, the year, the make and the model. You can't tell anymore. The first thing you did as a kid, when you were riding down the road with your parents, is you'd name every car.

I was the last shot in the bucket, OK? My nephew Carl's dad raced NASCAR-type races back in Riverhead and Islip, and all modifieds: '48s, '30s and stuff. Real cars. And Carl got into racing; I got into cars. My father, he was a mechanic almost his whole life. I've got a certificate on the wall of my office—dated 1936, signed by the vice president of operations for Oldsmobile—naming my dad the chief master mechanic for Oldsmobile. Also one when he worked in a Nash dealership. And then during the war he joined the military and went into the motor pool. Went in as a tech sergeant, came out as a tech sergeant. And was all over Europe, Africa, I mean all over the place.

It's kind of in the genes. But my dad never really worked a lot with me on the automotive stuff. I think because he wanted me to be better. You know, go to college; do better. So what do I do, at age 9? I carried an old 3.5hp Briggs and Stratton off an old reel mower and a bicycle frame to a local hot rodder's house. The guy's name was Ray VanZile. I was 9 and I think he was 21. Ray's probably my oldest friend. He's now in his 70s.

8. George Ross

Anyway, he had a T-bucket, and he had speed boats with Chevy motors in them. And I went up and knocked on his door and I said, "Mister, can you help me build a minibike?" He said, "Sure." He helped me build it. And when we got all done, I said, "I don't have much money." I reached in my pocket and pulled out a few bucks, because I was naïve, a 9-year-old kid, and he said, "No, cut my grass."

Not a problem. So all the way through, to the time I went to college, anything I needed. Because I did not have the tools. He had the welders. I'd use my dad's hand tools. One time we took a '62 Chevy II four door, we moved the engine back where the front seat used to be and we drove that with a straight axle. Ray helped me with the build and the welding. He would never take money. Even when I got to the point that I could afford to pay him. He'd say, "Cut my grass."

Now let's run the calendar forward about 30 years. He stopped at my house on his way up to Louisville for the Street Rod Nationals. He's never missed a Street Rod Nationals. Now I thought Ray was the coolest guy and I thought he built the neatest hot rods. But he'll tell you he built junk. He'd throw stuff together out of angle wire, thin wall tubing conduit, and build cars. He built this replica of a C-cab farm truck with a Chevy Chevette motor in it. He gets up there and the first day of cruising the fairgrounds, he

George Ross with his oldest friend and mentor Ray VanZine. George restored Ray's 1935 Pierce Arrow as a "thank-you" because Ray would never take any money for the help he provided when George was just starting out (photograph provided by George Ross).

blows the motor. Has a local garage drop the pan and they say, "No we can't do anything with it." So I go up with my trailer, pick it up and bring it back home and I rebuilt the motor for him.

After I get done rebuilding it, I am looking at it and this thing was just all flat black paint. I decided I am going to jazz it up. I am going to get Ray. I owe Ray some stuff. I paint the motor—what they call it—screaming yellow chrome candy and then I clear coated it and put silver metal flake in the clear coat. And put a couple of light coats on and this thing looked something like bling from the hood. You could hang it around your neck. He comes to pick it up, and he pulls out his wallet and he starts to pull money out and he says, "What do I owe you for the rebuild?" And I say, "No, no, no, no." Then I said, "You see that five-acre pasture over there? You see that '52 8n Ford tractor with the bush hog? You climb your ass on that and you cut my lawn. I've been waiting 35 or 40 years to tell you this." And so he did. He cut it and picked his car up. Ray was my mentor. He took me under his wing and got me started and it just progressed from there.

The very first thing that I ever painted was a Vespa motor scooter. I used my mother's Electrolux vacuum cleaner with the glass jar. Like a Mason jar. You screwed off the bottom. And I painted it with Mason paint, and it was just enamel. I didn't know what I was doing, but I thinned it out real well. It was a baby blue metallic, and I painted it in the carport of our house in Florida. I think that was the only time I never got a run. And now they are not called runs. They are called "professional flow indicators" in the business. Test panels; you know that you have enough paint on it when you get the professional flow indicator.

The best car I ever had? I would have to say the '56 Chevy convertible that I had. But I learned to drive on my dad's '55 Nash Rambler with the continental kit and all on the back. You remember that? He called that the blister, because it had some rust popping on it. But that's what I learned to drive on. It had a bed in it. The front seat folded down, and they always had the locks on the inside. Well, of course dad and mom are looking down, they're dead now, and they are smiling. I'll say I got my first piece of ass in the back seat of one of those. Actually it was a white rambler American, but that's another story for another type of book. But I guess it was a '56 Chevy convertible. I've had a lot of neat cars. But that's probably the one that I really wish I had back now. Because of the body style and all.

Of all the Tri-5 Chevy's, the '56 to me is ... well, I loved that. The one I liked the least is the '57. And I've restored several '57s for people. But I've just liked the '56, and I had a lot of good times in that car. Carl says the '56 was the ugly duckling of them all. Yes, that was the one nobody wanted. But I think the grill looked the best, the taillights ... just like '58 Chevys, nobody wanted that. Now everybody wants one. I had a '55 post car, six-cylinder, three speed on the column. I changed it to the floor; I was a kid, couldn't leave it alone.

Now my dad never came out and worked with me a lot or helped me or tried to get me into the cars. But he never stopped me. And he just said, "You don't want to do that. You are going to screw it." He let me learn on my own. But it was funny, he came home with this Vespa motor scooter when I was—I guess nine or ten—in the back seat or the trunk of the car, or wherever the hell he got it in. And he said, "If you can get it running, you can have it." That was his challenge. And in the next breath he said, "I don't have to worry about that because you'll never make a good mechanic because you're left-handed. Left-handers never make good mechanics." And my dad knew that if he ever told me that I couldn't do something, I had to prove that I could. And it was always, "No, don't tell me, let me do it my way."

That was my favorite saying: "Let me do it my way." So he kind of left it there. I'll never forget. It was the Allstate Vespa Piaggio sold by Sears. We went down to the Sears place, my mom got me a repair manual, and I started taking it apart. I'd save money cutting grass and she'd take me down and we'd buy parts. And I got it running by the time I was 10. I was driving around the backyard. At 11 I painted it, and I kept it. And I think I rode that to junior high school, instead of my bicycle. I kept it for a long time, but I ended up selling it when I got a car. It was a lot of fun.

But the '56 was my most favorite. My dad was the one that found it. We moved to Florida due to his health; he opened a bicycle shop there. It turned out to be Lou Gehrig's disease. He had it for a number of years; it wasn't a quick progression. He went into real estate, back when real estate wasn't making any money. So he found this, I guess he was over in St. Pete soliciting a house, and found this old man had this '56 Chevy convertible. Red and white interior, white exterior, white top. Gave $400 for it. Body and everything was perfect. It ran. It was just an old car. Just a used car. I messed with it, put tri-power, big cam, fuelie heads, built the motor into a 301. We took a 283, we bored her to 125 thousandths, made it a 4" bore, used 283 rods and crank. I went and got a set of pistons that I was told were 301 pistons. See at that time I was working, besides going to school. I was working at a gas station. Pumping gas and learning mechanical work.

So we built the motor, put it in, and man it revved high; just didn't seem to do so good. I came to find out that they were not 301 pistons. I always wondered why the pistons never came up to the top of the cylinder. But I was still at that time learning things. It did not dawn on me that there's something wrong here. So there was a hot rod shop in Tampa, where they built and raced a bunch of Willys stuff in the '60s. We took it down to them and they said, "Here's the problem, wrong pistons." It leaned out. They got the right pistons, did machine work and built it. I think I had tri-power on it, then from that I went to a tunnel ram, and I had an electric fuel pump along with the mechanical and it was not smart. I had it where you put the key in to turn the switch to prime the pump.

And I forgot to shut the switch off one day I was picking up my girlfriend. We were going to a wedding. I go to her house, naturally she was not ready, and I am sitting there waiting. We come out and the key was on, so it kept the fuel pump pushing and it filled the tunnel ram up with gasoline. And when it fired off it backfired, it caught on fire. I still didn't know it had a problem. We had the flush mount locks to the hood. Remember you had to put a key in and turn it to open them? I throw the hood open and guess what? Oxygen is coming from everywhere and it flared up. I had two little fire extinguishers in the trunk. I unloaded the two extinguishers on the motor. Nothing.

My girlfriend's dad comes running out with the garden hose. Now what happens when you have a fire going on gas and you hit it with water? It doesn't go out; it spreads the fire. It burned the whole engine compartment, melted all the aluminum on the tunnel ram and all, scorched the hood, scorched the cowl, cracked the windshield, and just started to burn the convertible top. Well, if I'd been smart, I'd have rebuilt the car. No, I pulled all the good stuff off of it; we pulled the engine and the Pontiac rear end and let a junkyard come haul it off. And that's when I built a '62 or '63 Chevy II. Guilt car. Should have never built it, but—you know—I knew everything. I was a teenager and I knew everything there was to know about, and my father was dumb. Amazing how smart he got in such a short time.

But a lot of the guys I grew up with back East got into cars. And in the last 10 years, my buddies Carl and Jack would come down here and we started hanging out and going to car shows. Wish we'd grown up together; we'd probably been in business now building cars together. Or jail, I don't know which one. So how I got into the restoration business? Well, let me kind of backtrack a few years. I ended up going into the military and I was stationed in England; I lived there four years. All during that time, I was into British cars. So that was how I got into British cars. That was the late '70s. I'd always played with cars, but it was always for myself, or I'd help a buddy. Got out of the military to save my first marriage. Should have stayed in and retired; but, you know, hindsight is 20/20.

After I got divorced, I was out of cars for about three or four years, getting my life back to together. I ended up getting remarried and we moved up here from Florida in about 1983 or 1984. Well, I told my wife I would build something for her. And before we moved from Florida, she fell in love with a kit car, the Gazelle, a replica of a 1929 Mercedes-Benz. There was a company down in Miami, Classic Motor Carriages, so I went down there and I bought the kit from them. And they said, "What are you going to put in it?" Well, I bought the one for a Ford Pinto motor and I said, "I was thinking about putting a V6 in or a V8." They said that's impossible. Well, I said I was going to put in an automatic, cruise control power steering, tilt wheel, make it a really nice car. Never happened; I didn't get a chance to build it. Still in the crate when we moved up here. So, the company that hired me ended up paying for two moving vans; one for my household stuff; the other for all my car stuff.

We built the car in Tennessee. We put the V6 in it, built it up to about 240 horse, had to take out the C4 automatic and have it a little bit stouter than what it was to match up with the power. Put in a posi rear, tilt, cruise, AC, everything on it that they said it could not do. We went down to see her mom in Ocala, we drove that car; my wife, myself, and my stepdaughter. We drove down to Miami one day just to show them the car and they still shook their heads at it, saying, "How'd you do that?" I said, "That's what you said couldn't be done." So that was the first. I got out of cars after that, I ended up selling that. Got a Club Cab Dually to haul my daughter's show horses. We got her into horses; we were into horses about five or six years. We got out of horses; smartest thing I ever did.

Then I told my wife, "Any car you want, I'll build for you." And I am thinking '32 Ford, T bucket, '40 Ford; but she saw an MGB going down the road one day and she said, "Gee that would be very nice." Well, OK. I cut my teeth on British cars. So we got one. Let's see, a year and a half later, $23,000 out of pocket and 1,000 man-hours, and we had an MGB with a 180 hp four cylinder, four-speed with overdrive, coilover suspension, House of Kolors Candy Brandywine over silver with Mini Flake in the last coat of the silver, and a custom interior. So we built the car, went to a couple of British car shows, just for the heck of it. And we ended up winning best of class and best of show. So next thing I know people are coming to me saying, "Can you work on my car?"

I actually built the shop and the paint booth in the basement of my house. The basement's about 1,400 square feet and it was really nice. And now I've got a 40 × 80 16-foot commercial metal building, we've got downdraft paint booth, mills, lathes, MIG, TIG, plasma, and a spot welder. We've got full fabrication facilities. And the business started in '93, when I turned it into a business. I've never done any advertising; it's just me. Now over the years I've probably had 15 different people working for me. Some of them lasted for a while; some of them didn't last very long. Because I am very picky about how I am going to do it. That was instilled in me by my father. You know, if you do it right the first time you do not have to go back and do it again. And give more than you are expected to give, and you'll always have customers. Right now we have about a two-year waiting list of cars to do. No advertising, everything is by word of mouth.

We've had cars that won at Amelia Island, Pebble Beach, Meadow Brooke, The Burn Foundation in Lehigh, Houston Classic, and Greenwich Concours d'Elegance. We've won 10 of the top awards at Concours d'Elegance. And here's the funny thing: I'm a street rodder, a hot rodder. That's my love. But people came to me because they saw this MG, when it was built as a hot rod, and wanted restorations. So we started doing restorations. And I still build hot rods, but I am only going to build it one way. There are only a handful of people that really can afford to build something to perfect as you can make it. The typical person doesn't have the resources to do it.

Now what has helped—for all of the builders—is the baby boomers. Me being one of them. Back in the '60s we looked at the cars and we drooled over them. Most of us we didn't have the pot of pee to throw at them. And now a lot of us have got to the point where we've got some disposable income. One of the first things that I'll tell a customer when he comes in, regardless of the car, is there's only a handful of cars that regardless of what it costs to restore, you're not going to be upside down in. But the typical car that somebody has that they want restored, they are going to put a whole lot more money into it than what they can turn around and sell it for the moment it's done. As soon as it's done, it starts deteriorating.

I'll ask them, "Why do you want to restore this car?" Well if the guy says it's sentimental, grandpa had one, it was my first car or I had my first sex in that car, I get it. I had that; I had a couple come in with a '58 Pontiac Chieftain, a four-door car. And I looked at it and I said, "Not mainstream; why do you want to restore it?" They said "We dated in this car in high school." So I said OK, so it's sentimental. And the wife goes, "You don't know how much sentimental." And the husband smiles. And she says, "I gave up my virginity to him in the backseat at a drive-in movie." And I said, "Do you want me to touch the upholstery in the back or leave it as it is?" I am thinking I don't know if I really want to get into that upholstery.

Then I've had people come in with an MGB. You can buy a driver that at least runs from anywhere from two to five thousand. We had somebody come in with a Plymouth Fury, and they said, "Well my grandmother bought one brand spanking new in '63 and gave it to me, OK?" That's one thing. Then you have the guy come who says, man, I found this '60 something Fury, I picked it up pretty reasonable, and I am going to put some money in it and I am going to turn around and sell it. And I'll go, "How much do you plan on putting into this vehicle? What do you mean when you say a little bit of money?" And invariably—I don't care who it is or what car—their numbers run between five to seven thousand. And they want to turn it around and make a quick buck. And I'll say, "Well, what do you expect for that money?" And then they say, "Oh, I figure we'll rebuild the motor and transmission, fix all the dents in the body and we're going to put a paint job on it." So I say, "You're talking about an every nut and bolt mechanical rebuild, what some would call a frame off, or stripped down cosmetic restoration. Well, I tell you what: Take that upper number, and multiply it by itself. So 49; that's how many thousand dollars, minimum, if the most strenuous thing you are going to do on the whole build is getting those writer's cramps filling out the checks."

Next I say, "Now take that $49,000, walk over to the crapper, lift the lid, put it in, pull the chain, and you'll get the same return on your investment." Or, "Turn around, and sell it. How much did you give for it at auction?" He says, "Oh, I give $5,000 for it." Then I tell him, "Well, put it back on auction. You'll get $2,500; take your losses and just move on. You'll be better off." And they look at you with this deer in the headlights look, like you are lying to them.

Now I've had some people put some money into a car that they are not going to get out of. One was a doctor here in Nashville. One of the best—well, how can I put this—I've never had customers, because we end up being friends. And I've only had one person that we parted on ill terms and he's been ostracized by every car builder and every car club around here. And he actually moved away because he was that kind of person. But when I am done with the car, we're all friends. And it's really worked out well that way. Customers typically end up as friends.

But back to the doctor. He is the head of oncology here at Baptist Hospital; he had a 1980 Audi coupe. And he comes to the shop, and he said he'd heard about us, and he wanted me to work on his Audi. I told him right away, "That's too new. We don't deal in anything like that." He kept coming back to the shop. Or calling, saying "Please come look at the car, please come look at the car." I didn't know who he was. At the time he was just some guy who walked into the shop. But I don't care who the guy is, if you are living under the Shelby St. Bridge, or you're Tim McGraw. So I told the doctor, "$75 an hour, for whatever number of hours, plus materials and parts. If you bring me the parts, not a problem. If I am going to spend time to hunt up the parts, I am not going to charge you the $75 an hour to find them. I'll make a 30% markup on them." That way everybody knows where you are at, and there's no surprises. Giving estimates is impossible to do. You don't do that with old cars. Everybody ends up getting hurt.

So we went to pick a car up, the '80 Audi coupe. I asked, "Why do you want me to restore an '80 coupe?" Well, he says, "It was the first car I bought; it saw me through college, medical school, residency, and Chicago." The lower eight inches of that car was rusted away. You could sit in the driver's seat and look at the road under you. And he held on to it. His house, in a gated community, had six garage doors. Behind five of them were cars, including a really hot BMW 7 Series, all black. And behind the very last door, the Audi Coupe. I looked at it and I said, "Doc, do you really want to spend money on that? I am not going to ask you the typical questions, it's clear this is sentimental. I am not going to insult you by asking if you have the disposable income so you don't take away from the groceries. But do you really want to do this?" He says, "Yes, I do." I said, "Ok, I'll do it for you. But I'll only do it one way." He said, "I know. I've seen your work."

We took it, we spent time, and he spent time trying to find body parts for it. We found one outer rocker panel; everything else we had to hand fabricate. We did it in black. It's gone to car shows and won at car shows, even being an Audi coupe. He's tickled to death. I may never do another car for him, but that car is more babied than any other car he's got. It gives you a good feeling in here when you see that the customer sees the car for the first time and that grin goes from ear to ear.

Clients gotta have a pretty deep pocketbook if they are serious about this. And they have to understand—maybe if they aren't car people they initially don't—that you don't know what you are getting into until you start blasting. What I tell the customer is that restoring an automobile or building a street rod from the ground up is not for those weak of heart, or light of pocketbook. Because there's a lot of emotional things that go on, that can put stress on someone. So I say weak heart because you are making decisions. You may think you want to do it this way, and then someone that really knows talks to you, "Well, here's why you don't want to do that." You've got to make the decision to maybe not do it the way you want to do it.

Or you get into it—like on the '70 Mustang we just had—get it blasted, and I get a call saying, "We found Bondo 4" thick. I have a rough hole the size of a half dollar and a guy had taken a piece of metal that was 3" bigger, laid it on it, just welded it ugly, and then put 3" of Bondo over it. Sculptured the car out of Bondo." And when you start seeing it, you go, "It's going to take this much time to do this, this much time to do that," and now your budget is blown.

That's why I say $75 an hour. It's the number of hours. And I'll be honest on the number of hours. And in a lot of cases—you can ask my wife who is a CFO at a company—I charge less than I should. That may make me an idiot but I was grown up poor, just a cinder block house that we grew up in. I look at a thing and I go, "You know, that took us 10 hours to do; we had a bit of a learning curve so we ran into some problems. It only should have taken eight." That is what I charge because I could not afford to have a car built by somebody. And the reason I know how to do all these things is because of that. If I wanted to have a nice car, I had to learn how to do it myself because I couldn't afford to pay for it. Maybe if I grew up with a lot of money, and money was no object, I'd just charge it up. I'd be rich and on TV and all this other stuff. But I am happy doing just what I am doing.

There are folks that get too big and in too deep. With cars pushed to the back of the shop and customers who see little progress on their cars. That's not a good situation. I have five people in the shop; one of them is a young man that is a high school kid. I've always got at least one of them that's got an interest in cars. I've always got one of them under the wing; they start off pushing a broom and then, "Here, bead blast this and then put it in the pressure washer cabinet" or whatever. Let's see, right now in the shop we've got a '29 Pierce Arrow sedan for mechanical work. I've got a '71 MGB GT that's all custom; the guy is going to have $70,000 to $80,000 in the car. When he is done, it's worth about $25,000. But it's got a blower, it's got every top-notch piece of equipment—just the motor itself is $9,000—he's chromed everything or replaced it with stainless steel. It's an ISCA show car that should sit on a pedestal with a trophy wife, front table. But he's going to drive it. We have another MGB GT that I had behind the shop that was mine. His buddy saw it and ended up buying it. I knew I'd never get around to it. It just came out of the paint booth and it's ready to start putting together. It also has a blower motor, but he is not going to do the chrome.

I just sent a '70 Mustang back to its owner. His job was changed and they cut his salary. This is a car where I've just finally said, "Look, this relationship is not good for you or me; I don't like having to chase you for money. You are now in a different position. You had champagne tastes and now a beer pocketbook. I don't like making enemies, and I can see it going that way, and I think that you need to take the car and go somewhere else with it." I've only had two of those. This job was going great; we'd just put the car in DP90 to keep it from flash rusting, and it's the one that when we got the Bondo off, there was another $15,000 of body panels needed to be cut out and changed.

We produce pretty high-end cars at the shop. The way I do the cars it's not necessarily that the customer wants it that way. A lot of them come in, they don't want a top show car; they want a really nice car to do good at local shows. In here, in my heart, I can't do that. But I'd build a car for myself that may look great, and I am going to drive it like I stole it. I am gonna rag it and beat it to death, you know? I missed a shift at 7,500 rpms in an MGB, I burned a hole in a piston, and I went home and rebuilt it. Hey, oh well? And I don't care what you drive; you're having a good time, it doesn't matter. You can drive your own piece of shit if you are having a good time that's great. In this thing, there's no such thing as wrong. There's some ugly, but no such thing as wrong.

So here I am in podunk Thompson's Station, Tennessee. Population 1,000. And I get to the level of Pebble Beach, or Amelia Island. I did not strive for that; it's just that the car turned out so good because that's just the way I am going to do a car. We started showing AACA [Antique Automobile Club of America] with the car and it ended up winning the AACA cup for that year, which is top restored car in the country for cars built from something like 1912 to 1925. I've also won the Bomgardner Award [for outstanding restoration of the year] which is from 1942 to present. Now all I need is a customer who wants me to build a car to compete for the President's Cup [outstanding restoration of the year for a 1921–1942 automobile], the only other top award I haven't won.

And we'll see what happens. Yeah, it's great. Pat on the back. It does not put any money in your pocket, maybe opens your hat size up one notch. We've taken a customer's car, a '59 MGA that actually beat out Ferraris in its class at Amelia Island. And who is that guy? The kid that is one of the top photographers of automobiles, Furman? Anyway, this car is in *Automobiles of the Chrome Era*. At the center of the book is a picture of this '58 MGA coupe that I did. And that's a great accolade, but does it put any money in your pocket? No, but at least when I die, maybe someone can show my grandchildren: "This is what your granddad did." That's, I guess, my mark in life. But we go to the shows and I would walk the class with the customer, and I'd go, "you should take second" or "You should win the class," or, "You should be third," or "We'll be lucky to get anything." And I would be honest with them. And whatever turned out, usually I was pretty right.

Right now we have this '54 Corvette that's on loan to the Corvette Museum. It won the Bomgardner Award this year. It won every venue, from hot rod shows, at Detroit and all over. We've won best restored in ISCA, it's won best of class in Concours d'Elegance, NCRS Top Flight, Bloomington Gold, Gold Spinner, and the Triple Crown. It has won everything it can win. Big deal. The only thing that makes me feel good is I know I did the best that I knew I could do. There is no such thing as a perfect restoration. I've never done a perfect paint job; I know where every flaw in the paint job is. But in the overall scheme of things, the cars are done well and as long as my customers are happy, I am happy. It doesn't bother me to go to a show and get beat because there's a lot of great builders out there. There's people that make me look like the shade tree guy. All I try to do is the best I can do on these cars.

You know, I've almost got to the point in my business that I don't take time to work on my own stuff because I got too busy. Besides that, during the day I work for a company that manufactures engineered equipment. I do intermingle that with the business, having to talk to customers on the phone, or my employees when they have questions, but when do I get my hands on? I do all the painting, I do a lot of the fabricating, and there's things that I engineer so I want to do it. Because it takes longer to show one of my guys what I want done. When everybody goes home in the evening, it may be 5 o'clock or 6, or it may be 10 o'clock at night when it's finally shut, then I can go down into the shop. Before I picked up my buddies for the [Nashville Goodguys] show I had to paint a quarter panel. None of the body shops wanted to touch it. This is for a big Caprice convertible. To get it done I may work until 2, 3, or 4 o'clock in the morning. I won't go to bed and the alarm still goes off the same time every morning, and I still get going. I've done that all my life: 18-hour days, 12-hour days.

Of course if I'd known I would live this long, I would probably have taken better care of myself. But that's just the way I am wired. I'll probably do that until the day they carry me out, boots first. Or they see me laying under the car and think I am working on something. Two days later they pull out the creeper and I am stiff as a board. But, you know, it's a passion, not a job. But I am starting to get to where I almost wish that a couple guys in the shop would take over running it. And let me be the consultant, or just paint. Let them just run the business. And let me do what I like doing and kind of back off a little bit and let me get off some of the pressures. It was a lot more fun back when it was simple.

So how do I do that? How do you do that these days when it seems you have to get bigger to stay afloat? Would I love to be bigger? Yes. But I don't want the headaches to go with it. My restoration business is a full-time business, but I have yet to write myself a paycheck out of it. My wife says, "All you are doing is keeping seven people employed." Well, if that's what the good Lord put me here for, serving that way, then so be it. As long as we're putting groceries on the table. I'll never be rich; I don't care if I am rich. I want to be happy.

I was in the corporate life, making more money than I am making now, but I constantly had those people standing on my shoulder, with the pressure. You know, "Gee, you helped us make 2 million this morning; what are you going to do this afternoon?" And I got tired of it. So I went to a different venue. The company that I work for probably doesn't pay so well for what I do, but we have no pressures as long as we do our job. It allows me also to do my passion.

Did I ever feel that, because it had become a business, that it diminished my passion? No. For me it's what I love to do. Ask my buddies. I've got a 1930 Graham-Paige automobile. The motor is having all the machine work done, so I can put it together. I've been gathering parts for over six years to build this car. I've got everything to build it. There are only two things I lack: time and the money. I really don't think I'll lose the passion; the only thing is that I am starting to get mentally and physically tired. I'll be

61 in a couple months. You know, at this time in your life you are supposed to be slowing down. But with the advent of computers, the advent of cell phone, beepers, pagers, all these different things, you don't have the down time that you used to have.

The last time my wife and I took a vacation was 1981. OK, I go to car shows; I guess you might say that's a bit of a vacation. When I take a customer's car to a car show, we are there for the customer so it's still business. I am still on the phone with my shop every day. But I really like going to car shows. Like my nephew Carl says, in this hobby you can see someone once a year or once every two years, you go talk to them, it's like you never left them. Like you were with them your whole life. It's like us here talking now. It's like we've known each other our whole lives. The connection is there and it never goes away. It's like we've been best friends forever.

And we are. It's a culture. We have something in common, whether it has two wheels or four wheels. We have something in common. And regardless whether you've got a million dollar car or one you built for $3,000 in your back yard, you've just got something you can talk about. Like I said, there is no wrong in building a car. It may be some ugly, and it may be some unsafe, but there's really no truly wrong. It's individual taste. And when you start looking out on the show field, you'll see everything and anything. It's just an absolute culture.

George Ross with the 1938 Cadillac restored by his shop. He has completed many award-winning restorations and has served as a Concours d' Elegance judge for over 25 years.

What happens to car culture in the next 10 or 20 years? From what I see—and I fly all over the country for my day job—the only place I see the young kids getting into it is Texas. Otherwise, you don't see too many young kids. Since they come out with the new Camaro and stuff, you see a little more. And because now NSRA [National Street Rod Association] has opened it up to 1980s, we all see a lot of grey-hair baby boomers at shows. These are older people who either wanted these cars when they were young, or had them. I had the car bug when I was young. I grew up with cars, I built cars. The boomers got the extra bucks in their pocket, and they do it.

I also belong to the AACA and I belong to British car clubs. The British car clubs are seeing young kids because they like the small sporty cars. And their parents or grandparents may have had them so some of the kids are getting into it. But mostly the X, Y, or Z generations—or whatever we got going—are into the rice-burners. This tuner cars thing is another culture itself. You know they have the "Hot Tuner Nights" and those things are big. The kids are all going into it. I think we'll evolve, because think about what we have now. At one time, the hot rod shows were all Model T's, stuff built in the 1940s, the original hot rods. You went to junk yards and built stuff. The Brass Era cars, all the people into that are all dying off or dead. Those cars are setting in museums; they are not coming to shows and being seen. So as things evolve, maybe we'll be seeing Mazda Miatas, or whatever.

Years ago, all you had was Chevys or Fords. I remember that Carl built a Plymouth. Nobody had that; there was maybe three in the whole family. Now, in the shows for the last six or seven years, they are all over. It's like trucks. In the last maybe 10 years you started seeing trucks all over. They are still available and a kid can afford a truck.

And survivor cars are coming out of the garages. We went to look at one yesterday, a '53 Dodge with a Gyro-Glide transmission in it. It's all there and could be restored. But to restore it back to original, it's kind of an orphan car. You are not going to find parts. But it would make a hell of a hot rod cruiser. Well, look at Steve Tracy, one of the sponsors at the Goodguys show who runs Advanced Plating. He's got the '40 Cadillac with the LS1 and he took it out on the autocross, he drove it like he stole it. And that's a high dollar car. And here you got the back end up, and the wheels smoking because it's not getting traction, and he's almost dragging the fenders. He had a hell of a time and the people loved it. There is no right or wrong. Go out and have fun. It may not be your taste in a car, but that's what his taste is, and he loves it. And you can appreciate the work that he's done. Maybe not your cup of tea, but you can appreciate the work that went into building it.

Also, some folks don't have a lot of money. They build what they can afford—a rat rod. And they have a good time with it, as long as they build it safe. I appreciate what they build. I like some of the stuff. I tell everybody I am going to build a rat rod, and they say, "No you can't; you won't allow yourself to do that." I've seen stuff like ... you've seen the little Dutch Boy statue? Well, one guy has taken one of those, put it on the radiator cap and drilled it to make it pee. And put a button on his dash that he can press to make it pee. His fuel feed was a cutting torch that came up from the back standing up, he put fittings in it, and he had his fuel line coming out of it, going to another thing. If you think about it, the ingenuity that goes into rat rods, it's like sculpture. It's art, it's automotive art. It may not be the prettiest, or even you might say, "It's ugly." But then when you look at what went into it—the ingenuity and the artistic talent—you look at a rat rod in a different way.

What would I build for myself? A rat rod Model A. My way. I want to build it with the blower and all, not for high horsepower, but build it for good low-end torque. It will sound good—have that blower whine—for driving on the street. I'll mold all the welds on the frame—just like I was going for a Ridler or ISCA. No body work, just the rolling chassis would be to perfection. I've already got the colors picked out; I am going to use Shimmering Silver, add Glamour Flake—the big flake like they used back in the day—and then PPG Hot Wheels Anti-Freeze Green candy. Paint the motor, frame, transmission, rear end, front end, even the wheels. I found 17" artillery wheels that are about 3½" wide on the front, and 5½" wide on the back. So big slicks and little front wheels. And right now I am having problems finding 17" wheel size of the old cheater slicks with the big white walls and the smaller tires with white walls.

Everything going on it is going to be highly polished, chrome, or painted. Headlights will be from a '37 Dodge; big around and long. They look like big long torpedo heads. I've got a Case tractor grille, a big round grille with louvers on it. I have not decided if I am going to put motorcycle fenders on the front and bobbed fenders on the back yet, only because in some states you have to have fenders if you plan to drive the car. And I plan to drive it.

It's going to be patinaed; I am gonna put "Ross Restoration Inc." on the door, and maybe a race number on the side. A patina-done body over a perfectly done frame. For the interior, I am going to have a really nice dash. It's going to be painted slick like the frame. And the upholstery? Remember back in the '50s and early '60s, they had the metal flake vinyl? Well, they have a green that's close to that anti-freeze green and they have the silver; we'll do a rolled and pleated interior. I'll build it as a finished rat rod that hopefully, if I do the detail good enough, I can take it to ISCA shows and be competitive.

That's the design right now. But my buddies are making me change my dream. They want me to use an already built flathead Merc with an overdrive transmission on it, with all the good parts, that I can buy for $3,700. A friend of mine has it; his business has dropped, so I'll buy it. I'll probably use that, put it into the car and then later go ahead and build the Hemi. And if I don't like the flathead, switch it out. But that's how I want to build the car. I don't know if it will come to that. I may drop dead tomorrow of a heart attack. But that's how I want to build one for myself as opposed to building one for a customer.

Am I going to find time to do this? Well, I am almost finished with my wife's car. I am building her a '52 Austin A40 Somerset, four-door version. As soon as that's done, I start on my Model A rat rod. I've got to finish something for myself. Eighteen years ago was the last time I built a car for myself. Like I said, I am taking time for myself. I've got to. It may be that I won't live long enough to do all the things that I've got to do. But I figure God gave me so many things to do in my life, I am so far behind that I'm going to live forever.

9

Lee Osborne (Penn Yan, NY)

Lee Osborne is a hot rodder with roots deep into racing. Before he could legally drive, he was racing a '34 Ford at a drag strip. Then he got into dirt track racing. Recently he was elected into the Dirt Track Hall of Fame. Oz is Old School. And a minimalist. He builds traditional hot rods in his home shop. According to Lee, a hot rod just needs good steering, good brakes and a good radiator. As he puts it, "The thing doesn't overheat, goes down the road straight, and stops when you want to stop." The rest of it? He's OK with no doors and a pop crate for a seat. Like I said: a minimalist.

LEE OSBORNE: I actually had my first lot car when I was 11 years old. I grew up on a farm near Spencerport, New York. It's about 15 miles from Rochester. There were about a thousand people in the town. And there used to be a track at the Monroe County Fairgrounds, which is next to Rochester. We used to go to the fairgrounds and watch; all year long they ran half mile dirt track there, in the old coupes.

I hung around the farm, learned how to drive almost before I could walk. Anything with a motor on it I wanted to drive. When we were kids, everything we did was the Joey Chitwood Thrill Show. We'd build hay bale deals and put guys on them and drive through them. We did all the thrill show stuff and everything. Mostly stayed on all four wheels. We had one we tipped over several times. Because it was fun being in the ditch.

But I always had an interest in speed. I had lot of cars. But as I got older, I just knew that I wanted to race. In fact, I was halfway through my junior year in high school and I got out of school. I was racing motorcycles, scrambles at the time, and working two jobs, so I could do it. I finally went to work for Turner's Excavating Company. I knew them forever, because I used to hang around there. They had modifieds that they ran at Monroe County Fairgrounds. All us kids that hung around there, they put up with us. We were all a pain in the ass.

And eventually they had an old wrecked car out back that Bobby Hudson had crashed. It was wrecked on both ends but the cage and the body and everything was good. I asked what he was going to do with it. He said, "Nothing. Why?" And I told him that I wanted to race it. He said, "Everybody wants to race." I said, "No, I want to race, I want to do it." He said, "Well, you take it, put in the shop, cut the front off, and you splice a front clip on it." He told me what to get, where to go to the junkyard, the deal, you know. Because I had been there with them and I knew half-ass what I was doing. They told me where to put everything.

It took a year. I finished it for the next season. As I was working on it in the shop, the guys that ran the equipment at Turner's said, "You going to get that thing done? 'Cause there's a crusher in the old barn right across the road from the shop." And I said, "Yeah, I'll get it done. I'm spending money as I get it, you know I am out of money all the time." Well, they bought me a camshaft and my first set of tires and a set of cylinder heads. And everybody said, "You gotta beat Ray Turner's (Tubby's) car. You better beat that car." I said, "Don't worry; I'll beat him." Well, anyway, I'm 18 at that time. So I got it done and we went to Shangri-La Speedway, down by Owego, New York. And first night, I didn't have a trailer or a tow truck or anything; I had a jack, an air tank, and—I think—one or two spare tires. Tubby was flat towing it. We put street tires on it and we towed it behind his hauler. Blew the top of the radiator going up the hill out of Ithaca and had to go to a junk yard to get another radiator and put it in the truck, you know, the whole deal.

Lee Osborne at his home garage rod shop showing western New York speedway programs from early in his dirt tracking career (July 2009, Keuka Park, New York).

Well, we ended up running third in the heat that night and fourth in the feature. First night I ever raced. How I did that I don't know. It's just all heart. It's what you want to do, and if you want to do it bad enough, you can do it. That's what it amounts to.

Before that I'd raced motorcycles for two years so I had a feel for loose and tight. And I had the ass to brain cable; I kinda had that feel. And this was pavement, I started out on pavement. This wasn't dirt. We ended up winning—I think—three races that year. By the second week I thought if I could run fourth the first week, I can win the next week. You know, I'm 19, thinking I can really do this.

So we went down, I spun out, bent the rear end housing, and hit the wall, did that about two or three weeks in a row trying to go faster than I was capable of going. An old guy named Fran—he had a trucking company and owned Shangri-La Speedway at the time—pulled me over and he says, "I gotta talk to you for a minute." We went off to the side there and he says, "I'll give you a little advice." And he said, "You think about this. If you slow down, you're going to go faster. Just think about what I am telling you." I said, "yeah, OK, OK." And I go out. And he was right. I just had to be a little more consistent. And figure it out rather than just crashing around.

I figured it out probably halfway through my first year. You know, you can't get to the front in two laps; you have 25 to do it. We ended up winning, I think, three or four races the first year. And a lot of top fives. So, next year we built a new car. I paid for it by working at Turner's. I was working at a Chrysler garage when I built the first car. And Ray Turner, he came out one day and he said, "You know, Dick needs some help in the shop. I want you to come over and talk to him." That's when I went to work for them, on the equipment and stuff. And the second car, I was working at Turner's full time. They hired me halfway through that first car.

And I made good money, for a kid with no education or anything. I worked a lot of hours, but they were good to me. They were like my brothers, the whole bunch; they helped us a lot. And two or three other guys that got to race because of them. They helped them a lot, too. One of the guys has the driving school in Florida now, does a lot of the NASCAR drivers.

The second year I was either going to go to Vietnam, or join the National Guard. This would have been '66. I joined the Guard and I had to go to basic training. I went through that winter and then part of that summer. I flew back home, didn't get to race a lot. But we won a few races, but we built a sedan, with the motor way back in it, our own ideas, instead of using a coupe. We did this thing. Well it was fast, but it was hard on tires. Because we had too much rear percentage in it. We were just kids and we didn't know. We had gone to Owego for years and watched those cars that had a lot of offset. And we are sitting there that fall, and I was out of basic training, went home and everything, and I said, "Well, here's a common sense approach to this." I said, "If the right rear is getting too much heat, let's offset the motor."

So we offset the motor to the left four inches. Left it in the same place. Offset it, built headers and the whole deal. We won 29 races that year, our third year, 1967. And then the next year, Dutch—my father-in-law now—quit driving for Turner's, to build his own car; he wanted to race his own car. So he quit driving their car, and they hired me to drive their car. That would have been 1968. Then in that year we ended up, I don't know, we won 18 races and blew up every motor they had. But the Turners both got married and they had little kids, and they weren't spending the money on it. And I was frustrated. And that's the year I went sprint car racing: 1969. I bought a car and went sprint car racing. I moved to Pennsylvania, lived there six years. And then I ended up in Indiana for 23 years.

I moved to Indiana because that's in the middle of everything. We traveled, we ran up and down there. When I lived in Pennsylvania, I ran there for four years and then I started to run USAC non-wing open wheel racing, which ran in the Midwest. I drove back and forth every weekend. And ended up having a bunch of buddies out there around Brownsberg, Indiana, so I ended up moving out there, to the little town of Jamestown.

That's where we started building hot rods. I raced up until '84, and ran all sprint car stuff. Ran World of Outlaws, All Stars, USAC, IMCA, and everything. Anyway, all this time I always had hot rods. I've got a picture of a '34 Ford coupe that I had in '62. My first hotrod. I had a Model A before that, but I didn't get that done. My '34 Ford had a flathead; we put a 348 Chevy in it. And I didn't have a driver's license at the time, but I drag raced it at Spencer Speedway [Williamson, New York]. A buddy used to tow me up there, and we'd drag race it. No license; I was 15.

And then I ended up getting away from that because of the roundy-roundy stuff. And then I quit in '84 after 20 years racing. I quit because it kind of outgrew an individual. I did it as an individual; I built all my own cars and made cars for other people. I assembled all my own motors, I had the machine work done, but I put them together. And I towed it, I fixed it, and I did everything to it. I worked seven days a week, every day of the year except Christmas—10, 12, 15, or 18 hours a day. Whatever it took, or all night; it didn't matter.

Whatever it took. I raced. I tried driving other guys' cars in Pennsylvania, and it just wasn't good for me. I understood the chassis real well, and when they did something I didn't think we ought to be doing—whether it was better or not—I was not happy. Maybe it was better—but racing is a funny deal; it's about 80 percent mental. And when something isn't happening on the race track, you can't sit in the car and blame the car. On dirt you gotta try to figure out a way to at least make enough money to get to the next race. You try to make stuff happen. And that's why it's so much fun. On pavement it's cut

Lee Osborne replacing a brake light switch on Jim Hawver's '32 Ford built by Lee 17 years ago. Over the years Lee has built and sold more than 100 hot rods, by his estimation (May 2017, Keuka Park, New York).

and dried; you put a setup in the car for a certain racetrack and you run it that way every night when you go back. Might change a pound of air, or whatever jet the motor's on.

Dirt racing is a whole different deal. Nothing's the same, ever. It's mainly the racetrack. You can go to the same track and it will be five different ways, five different nights. Because it's March, or it's June or it's 80 degrees, or 40 degrees, it rained, or it hasn't rained, they've worked the track, they didn't work the track, they put new clay on it, or they didn't do a thing to it. There are a million things to adjust for.

The cars are totally adjustable. It takes a long time to learn how to dial it in. You write everything down, keep a lot of notes every night. Check, weigh the car, measure the car, do all that stuff. And write down what you did, and what was wrong. You end up with a thick book of stuff not to do. And a few pages of what to do. And there is no right or wrong. If you gotta put a square tire on to win, you just do it. You don't question it. You just do it. And then you take it back off and go back to your basic setup for the next race.

What did I learn from this that I build into my hot rods today? Well, to just keep it simple. People think you need all this stuff you don't need—air, an independent front end, and all that kind of stuff—you don't need any of that. What you need is good steering, you need good brakes, and you need a good radiator. The thing doesn't overheat, goes down the road straight, and stops when you want to stop. The rest of it? You can have no doors, pop crate for a seat, and long as you got light at night I guess you are good. And other than that, those are the only necessities.

It's a kind of fuel, air, and spark design philosophy. Yeah, that's where hot rodding came from. Originally, to build a hot rod, guys got a car and they took everything you could take off to make it as light as you could to go faster. Right? And that's the same way today. I came from the race car side of hot rodding. A lot of people bought a new, whatever, Chevy, in whatever year, and they waxed it every day and took care of it. Then they bought another, and they'd say, "Boy, I'd like to have an old car." Then they'd buy a hot rod. And they are intimidated by everybody unless they've got heat, air, cruise, tilt, tinted windows, a $20,000 paint job, polished wheels, stainless exhaust. They feel the minute they pull in they are being judged. So they are from the "show type" side. Well, I'm from the other side, the "I don't care side." As long as it runs, it steers, and makes a lot of noise, and all that, that's where I come from. That's what is good about all this; there's something for everybody. You can be here, or there, and it doesn't matter. You can park together and have a good time. That's why it's good.

And that's why Syracuse [Syracuse Nationals Car Show] and all these places is such a success. Because it's that way. There's room for everybody. And once you are in it, and once you figure it out, once you've been in it long enough to figure out that it doesn't matter, you can go, you are accepted whatever you got. Until then, you are all uptight. You are one of those, "Don't touch my car" guys.

And those rat rods? Well, they are taking that to an extreme now. Which is good because there are young guys in it. Or it will die. It did in the '60s. It died because the muscle cars came out and hot rods ended. Because for three grand you could go twice as fast. And I did. I had a brand new Plymouth—a brand new '65 Plymouth, 426 wedge, four speed, the whole deal. Anyway, the young guys now are into the ink and all the whatever. And so what; that's their deal.

If you go and park in the rat rod corner you're an outsider if you don't have all that stuff. I was doing this when they were a dream, you know. They weren't even a dream. I remember it when—back in the '50s. I was born in '45. I've been around it a while. I was around cars from the time I could say the word. That's what I did. My dad was a Ford mechanic early on. I've seen my share of the whole deal: the racing side of it, the hot rod side of it, the whole thing.

And it's good that young guys are involved, but as far as all this drama that they go through, hot rodding wasn't like that back in the day. Guys had primer because they didn't have enough money to paint a car. It wasn't because that's cool; it's just because that's what it was. And they were built out of the junkyard because there was nothing else available. Our first race cars were built out of the junkyard because there's no speed parts. Today if you need a quick-change rear end you order it. The old stock cars and hot rods, I mean, they were all old Ford parts. My cars are still that way. I did all the power stuff; I did all the cars with all the glitz, no chrome, all the polish and everything. And every time someone got one of those cars done they come back in a month, and they say, "Jeez, I am sick of cleaning and waxing that thing. Next car I do it's going to be the simplest car, and then you just enjoy it." You know, you figure out that that's all it takes. It takes a hot rod.

What is a hot rod? Well, the new cars are not hot rods. Because they are more muscle cars. They call anything a hot rod now. But I would say it's your interpretation of a '30s to '50s car. Your own interpretation of it, something that's loud, obnoxious, and fun to drive, you know. Really they were a crude race car when they were built originally. That's what they were supposed to be. So what they are is just a crude version of a race car.

Some rat rods today I really would not feel safe in. But that really isn't a hot rod. That's part of that culture deal. The more outrageous the better. And hot rods were not like that during the day. I mean they didn't have truck wheels on them or their rear-end dragging the street and all that stuff. They were going to use them, drive them, drag race them; whatever you did, they still had to be functional. Those aren't functional. That's their interpretation of what one is, you know? And that's OK, but it's not really the way they were. They weren't like that. Nothing I remember.

I am building '50s and '60s style cars. The way they were. And I am not much into '30s stuff because I don't like flatheads. Not that I don't like them, but I want to go and come home. I don't want to work on them all the time. And they are fine if you want to spend 5, 6, or 8 grand on a flathead that's got 150 or 200 horse. If you are lucky. And gotta work on old transmissions. Well, we put one in Gary Buehler's car. And we had it out once and now we have to go through it again. There's nothing wrong with them, but you gotta work on this stuff. And the rear end is the same deal. You put the early rear ends in them and they shear the keys. And they leak, see? And that's fine, but you gotta work on this stuff. That's what you do, you work on them. And I don't like that.

I work on stuff all the time. I just want to turn the key and spin the tires and go. But it's like everything; it's changing. You know, I think the biggest change is they have these big events and everybody gets screwed up on what it's really all about. Out in Indiana, when I lived in Indiana, there was a lot of little clubs that had been in these towns—I mean there's a lot of little towns out there with 600 people in them—well, there's six or eight guys in that town that all had hot cars. What else are you going to do out in the middle of Indiana? They are all farmers; they had nothing to do Friday or Saturday night, so they went to town—maybe to Indianapolis or to Lafayette or somewhere. You went to the biggest place around where all the broads were, you know, all 12 of them. Whatever. And hung out. You had to be cool, you know, so you had to have a hot car. And that's what you spent your money on; there was nothing else. That was the way Spencerport was, 15 miles from Rochester. But we weren't city guys; we were hillbillies, really. And everybody had some kind of hot car, had loud pipes or you had something, to carry your ego, you know.

All those little towns had a club, had a jacket, a plate, or whatever. They have a hog roast or a chili deal, or a whatever. We had one for 15 years at my old shop. And they still have it. We have a lot of people now, started out small, but those are the things that are fun. Because it's not overwhelming, everybody drives their car, and you don't have to clean them up. There are no awards, there's nothing for nothing, so you drink beer and everybody has a good time and everybody goes home, and hopefully you don't crash. And that's how it is. And that's kinda what hot rodding is. And all my buddies out there, and the guys that come back here, they all just drive the shit out of their cars. They got cars with 70, 80, 90,000 miles on them. And that's what they drive.

When I came back here, for the first two or three years, that's all I drove. All I had was hot rods. Well, I had an old shop truck but I didn't drive anything else. And back here it's different. You have a short season, and anything I owned, that I had back here, somebody wanted. When you are in the hot rod business, and somebody wants to give you more than it's worth, you sell it. So I've sold everything. I started two cars last winter and didn't get either one done because I sold them before they were done. So now I am working on a five window. I am going to finish that car, because I want a five window; I want to get one done.

Why a five window? Ahhh…. I just always liked them. I actually like those and Model A coupes. I've had a million '34s and '33s, and '32s and roadsters and all the stuff that you are supposed to die for, you know? And I like a '29 roadster; that's probably my favorite roadster body on a '32 frame because of the little cowl and the curve in the quarter panels, and the swoop off the drivers surround and all that. They got a real nice shape to them. Or a '27 T. I had a '27 T with a flathead that I drove for seven years. It had a V8 60 in it, and a Columbia two-speed. No power and it smoked. But it was fun to drive. And that or a five window, a chopped five window. Or a Model A coupe on '32 rails. I remember that stuff, so it's what I'd like. Three windows were always great but they were big money back in the '60s. There wasn't that many of them and everybody had to have a '32. There's nothing wrong with them; they are a nice car.

With enough money, you can buy a new steel body from Brookville Roadsters near Dayton, Ohio. I had a body brought to me at Carlisle this spring. We see them at Carlisle every year. They do nice stuff. Their stuff is excellent. It's the best stuff going, and there are a lot of guys doing it. Their stuff is good for this as a hobby, you know? Because very few people can resurrect the junk. It's hard to. But I don't buy the old metal anymore because I don't have the ambition. You know, we did so much; we built the bottom half of them before there was door skins and all the patches. We made all that stuff. We had a wheel and all that. We used to make every panel, because you couldn't buy any.

Nowadays if I do two complete cars a year that's a lot. I do a lot of chassis stuff, and

Lee Osborne's chopped '29 Model A five-window coupe; sitting on '32 rails, currently sporting a 327 with a Muncie four-speed transmission. This one he is not going to sell (May 2017, Keuka Park, New York).

some top chops. After I chopped one '32 the owner came in and wanted me to fill the top. And I wouldn't fill the top; it's a brand new body. I said, "You don't want to fill the top." I said, "I'll make you a steel insert to put underneath and we'll put a soft top over it. But don't fill the top of that brand new car; it'll look like every other glass car out there." I finally talked him out of it. In fact, I had a sheet of steel over there and I was going to start wheeling out the top. And now what we're going to do is put the wood back in it. And finish it with a steel top that you can put in with fabric over it. He saw this in a hot rod magazine. In fact, I've done a car like that once because its top was so messed up. But that's what I think we're doing, so at least the look is there. So I gotta do that.

Projects have a way of expanding. I sent it to my paint guy to put it in primer because it was rusting. He didn't do any bodywork on it. It's just the way it was welded back together. It came for a top chop, and I ended up doing the whole chassis for him. That's kind of how it goes. You think you are going to start on something new. But between that chopped job, and the chassis, a new gas tank and a battery to put in it, I got my work cut out for me. And I don't want to start my next project till this one's out of here. And this one is a week or two from being out of here. This guy's going to finish it. I want to get this done. Then I can take the next project, sort through an old pile of parts and spread it all out, paint the frame and do whatever I got to do to. But you get too much stuff lying around and you can't get anything done.

Every job evolves into more than you've planned on. It's the "while it's here" factor. They say, "Well, while you got it there you might as well...." Anyway, so what you think is a month worth of work turns into three months, or whatever. There's only so many of them a year, you know? But it's fun. I haven't had a hot rod in two or three years. I built a little '34 Ford tub that's got the red wheels on it, that's got an old 292 Chev, dual quads, and a four-speed. And the guy who bought it from me drives it on the street. He also bought a '27 roadster that I built. And then I had a '33 Ford coupe that a guy from Waterloo bought. I built that tub and I drove it 500 miles.

I haven't had anything to drive, so this year I thought, well, I am either going to buy a Corvette, or a year-old GT Mustang, or something. About two months ago they had a deal at Knoxville, where I got inducted into this Sprint Car Hall of Fame. I didn't have anything good enough to drive out there that got good mileage, so I rented a Toyota Prius. I go over to Geneva and pick this thing up. I didn't know what it was; I had told them that I wanted a medium-sized car. A lady showed it to me and I said, "That's mid-size?" And she told me that, "Oh yeah, we got smaller sized stuff than that." So I jump in this thing and I drive to Knoxville and back, 80 miles an hour, and I get something like 30 miles per gallon, and it's all right. It gets me there and gets me home, you know. And all my buddies at the induction ceremony are all laughing.

After that, I was up at Turner Automotive [Victor, New York] one day and I saw this black Corvette. Black wheels, black interior, had the red brake calipers and all. And the wheels: Normally the wheels are all polished; this had them dulled out and they were black and I thought, "Man, that thing's really nice." I said to myself, "I ought to have that." I know Turner pretty good; he's the guy that sells all the Corvettes. I said, "What's the deal on that?" And he comes over and says, "Guy had it, lost all his money. I bought it for a song. If you want it, I can give you a hell of a deal on it. I'd sell you that cheap just to get you in one." So he told me what he wanted for it. It was $15,000 less than it was worth, you know? I thought that'd be fun, so I said, "Let me go home and tell my wife what I am doing here. I gotta think about this stuff for a day."

I got home, and told my wife that I'd stopped at Turner's and they got a nice Corvette. She asked, "What year?" I told her 2008. And I told her it was black, with a black interior. She said, "Black with a black interior?" Then she said, "Well, are you going to wax the outside and your dog Waldo is going to clean the interior, or what?" She knows I don't wash anything. Waldo goes with me everywhere. He's taken two interiors out of the Tahoe. Visors, seats, door panels, headliners and everything. He's finishing up on his second interior. And I just had my shop truck's seat redone. It's just what we do. And the same with my hot rods, he rides in them and eats dog food in there, the whole deal.

My wife starts laughing. And she says, "You buy that car and you'd be bored with that in a week." And I said, "No, I am telling you...." And she is laughing so hard that tears are coming to her eyes. I called up Turner up and told him that, "She is still laughing about me having a black car with a black interior and cleaning it." Well, he said, "She's probably right, I know you." I said that I would probably build another primered hot rod, and he said, "Well that's probably more your forte."

So anyway, I went by Chrysler garage, and they had one of those Ford Bullet Mustangs. You know, spoiler, green, and wheels are dark on them, and the whole deal. A used one, 2007. I thought, "Jeez, that thing looks good. So I stopped and was looking at it and you know those salesmen, right now they are like turkey buzzards, and they are right on you. Some lady came out and she said, "You like that car?" And I said, "Oh yeah, I like that car." Then she said, "It can be yours." I said, "Well I am sure it can." Then she said, "We've got a thousand dollar deposit on it, so we can't really do anything on it right now till this guy says yea or nay." Well that's good, that's where I want to be. So I left. I came home, and this kid shows up here with it. He calls me up, this dealership kid. He says "Do you want to drive that car?" I said, "Nah, I don't need to drive it." He says, "Hey, I'll bring it up." So he came up and I drove it. And you know what I thought about it when I was driving that Mustang? From the outside they look cool. I got inside that car and I drove it. And I thought, "This is no different than that Toyota Prius." On the inside, it's the same thing; it's got a modern steering wheel, modern dash, modern seats, modern radio, modern air, modern everything. And my wife was right. In a hundred miles, you are bored.

It's different when you drive a hot rod. I don't care what you do, what it's like, whatever. When you go out and somebody's going by and they are smiling. And you are driving and thinking, "Things aren't bad." You can be having the worst day in the world and it's just that better. And they are always fun to drive. That's what hot rodding is.

People buying these things, they go and buy a used hot rod, and then they want something built. And if it's a guy that doesn't really know much about this, I tell them to just go to about five shows, look around, and find out what you want. You'll find a used one cheaper than you can build one. Make sure you want one, make sure you like them. That's the best way to get started. Rather than having something built, and getting in it three times more than you want in it, and not liking it. And a lot of people just don't. It's like a Harley. Lot of guys buy a Harley and later say, "Why the hell did I buy this thing for?" Wind and rain, or your skin burns up, but if you like it, it doesn't matter. You don't care.

That's the way these things are. That's the way a roadster is in June; you got no top, the sun's beating down, the wind is in your face, and your ears are flapping. And you've got earaches. But if you like driving your hot rod, it doesn't matter. It's what you do. I just think they are fun. But I have not had one for a while. That was my reason for wanting a Corvette: to have something fun to drive. And they are not fun to drive. So I am glad I went through all that and decided to build a hot rod for myself.

My next deal was, "I am going to buy a '32 to '34 Ford." I haven't done one car yet that I wouldn't get in and drive to California. I haven't done a car that I didn't think would make it. My '32s and '33s, I had those all on the road at one time. And my T roadster. I've had 50 hot rods and I'd jump in any of them and go to California.

What's in our future? Well, I've driven a hybrid and they didn't convert me. I can tell you that. Is that the future? Nah. I mean, the old cars that are around now, there's a lot of guys that got 25 and 30 cars hoarded up. They are going to die. And when they die, there are 25 or 30 more cars that are going to be out somewhere. After they tip over their family won't care about them. They want the money. Because they want to go and buy a new BMW, or whatever, you know?

The traditional cars are always going to be around. And to me—traditional hot rods, that's what it was founded on: the old dry lakes cars. Or the old drag cars or the old roadster roundy-round cars and the coupes; that's how it started. And those are always going to be in vogue. And there's always going to be somebody that passes on that memory to somebody else, of that type of car. All the trends come and go. Like all the stuff we built with no chrome on it, and flip out headlights, and air conditioning, and all that stuff. That was a trend. Then the fat-fendered cars were big for a while there. Everybody had to have a fat-fendered car, you know? They were a trend. I mean, I don't even know all of them. Whatever it took to make money, that's what we built. And now, I just won't do them. I send them over to my buddy Birosh. He'll work on anything. He don't care, you know? I just don't want to do them.

But I think these cars, as you go down the road, they'll always be around. If you think back to the '40 or '50s, this is what I remember a hot rod being. You know if you went to the Dog 'n Suds or Ackerman's, or the restaurants that were around here at that time, drive-ins, Carroll's Restaurant, 15-cent hamburgers and stuff. Everybody back then was either with a '55 Chevy or a '54 Chevy or a '51 Ford with an Olds motor in it, or a hot rod. And that's always going to be around. And unless the government says you absolutely cannot drive a V8 car. And that's what—if we keep these guys in there now—that's probably what it's going to come to. Which is stupid, because it's such a small percentage of the total problem. But who knows. I hope it never gets any worse than it is now.

What's going to happen when we lose the old guys with the kinds of skills to do work like I do in my shop? Well, that's going away. But right now there's a big interest—well, not a big—but because of TV, there's an interest in metal shaping. You know Jessie James, you see him there, making the side of a gas tank on a sand bag, with a planishing hammer. And he's got half the tank done in 15 minutes. But they are not showing you the other half the day it took to make that piece wrong twice and then to get it right. So they oversimplify everything. So everybody now wants an English wheel and planishing hammer, a mallet and a set of dollies. And they are going to build a Harley gas tank or build a '32 Ford body for one of these mail order '32 chassis.

Well, it isn't gonna happen, you know? It's like me being a brain surgeon. I mean they probably can't do it. Because their head's not in it. And it's trendy right now to do that. Maybe you can say, "OK. I built that ash tray in my car." Well, this stuff—all of it, any of your sheet metal work—is just blacksmith work. And it's just lighter metal. That's all it is. And there's no interest in really learning that. I mean people just don't care about it. There are a few guys right now that do it because it's stylish, but in reality, there won't be anybody doing it, you know? Everything will be composites or fiberglass, or whatever.

Well, hot rodders—like the guys that started racing years ago—they went to the junkyard, like I did. We started with a piece of junk. We cut the ends off, put pieces on it; when we got all done, we knew what we had. I knew why I put this here, what this is going to do, and everything. You learned from the ground up. You learned by experience how to make it. It's not like that now; you can buy every piece. It's all 1-800 stuff. Back in the day we dug it out, steered it home, took it all apart, and got the frame. You try to weld something and you couldn't weld, you know? But some guy said, "Well, I'll weld it for you." And you got the thing going. And you were proud of it because you built it and you found all the pieces and it's what you wanted.

Or maybe it wasn't what you wanted but it was what you could afford. And at least you were in business. At least you knew why you did it and your heart was in it, you know? It's not like that anymore. People will go to one car show and they say, "Gee, that's kind of cool. Let's sell the Porsche and get this and do this." And the guy gets his car and he's standing there at the gas station saying "Can I ask you something? How do I put the gas in this thing. Where's the gas filler?"

Oh well. Maybe that's what is good about the car hobby. There's something for everybody and it's all OK. But if you ask me what I do or what I like? Well, I do what I like. And it's fine what everybody else likes, you know? Then everybody says, "You ought to...." And I say, "Well, I really don't want to go to that because it's not my deal."

I'm not into car shows. Just local get-togethers and cruises. Just five guys get together and say, "Come on over. We're getting a keg of beer and cook a deep fried turkey or something, you know? Cruising around with these things, that's what it's all about. Use your car. Not sitting in a lawn chair, you know? Not to get a trophy; it's about driving your car. Driving it there, driving it while you are there, and driving it home. You have some kind of gymkhana thing or whatever they call it, you know? Have drag races, burnout contests, and all that kind of stuff. Stuff where you can use your car.

The big shows have turned into vendor events. You got all this shit up every place you are supposed to park. Put the vendors out in the parking lot out by the road. Let all the cars come up here in the shade under the trees, in the grass and all that, where they can enjoy 'em. All the car events turn into a deal for vendors to make money.

You ask how all this stays fun? Well, I don't know if it is fun. But it beats having a job. It's a lot of work. You know the worst part about doing one of these things is if you are a painter. If you had to block sand one of these things ten times to get it nice. And then start on one next week. I don't know how they do it. What I do goes quick. For painting, there are the fumes, and it is labor intensive; it's all hand work. Some say you buy a machine and sand it. You don't. If you want it nice. It's awful. But painters do it. We got a kid right here—a guy named John Carroll—he's great, he does one then he does a bunch of trucks or some other kind of work. Gets away from it; then he'll do another one. He won't do two in a row. He just can't. But he makes good money doing it. I do everything but paint and interiors. What do I look for in a painter or an interior guy? Anybody you don't have to fight with. That's it.

I am here in the Finger Lakes area building hot rods the way I want. Back to the roots of my hot rodding. I remember going to Monroe County Fairgrounds, Spencer Speedway, Niagara Raceway Park, Erie Dragstrip, Dunkirk, Savannah. I remember all the small, East coast-style dragstrips and oval tracks that there were back then. And they used to run in the '40s. I don't remember this; but I remember the cars. I never went to the races then. But in the '40s they ran midgets seven nights a week out here, after the war.

Everybody came home from the war and they had to do something. They all had midgets in New York here. They ran them at Civic Stadium, I think, two nights a week in Buffalo and right on Scottsville Road, right up by the airport. A little later, we raced indoors at the Rochester War Memorial. We ran Daytona twice, ran the Mint 400 in Los Vegas in a dune buggy, we ran stock cars, we ran all the sprint car stuff, we drag raced and had hot rods. Never did race a tractor.

But I had a great life. Perfect. Couldn't do better. Grew up in Spencerport, the best possible place at the best possible time that you could have ever done it. Had all the opportunities in the world that you needed. And then I got to race for 20 years. I am still involved with race cars; I went last night, but we got rained out. And I get to build hot rods till I tip over. I mean what else is there?

10

Darryl Starbird (Afton, OK)

I met Darryl Starbird in June 2012 at his National Rod & Custom Car Hall of Fame Museum in Afton, Oklahoma. I'd been there 10 days, participating in the chop of my '58. Darryl's grandson, Dakota Wentz, did the work. I was Dakota's helper; we worked for long days in the heat. Most of the time Darryl was beside us, working on his newest project: a highly modified 1957 Cadillac Brougham called the Shark. Darryl occasionally lent his expertise to our work, but Dakota knew all the tricks, having worked beside his grandfather for many years. As I reflect on the experience, it is clear to me that Darryl has passed a wealth of knowledge to Dakota, an outstanding craftsman in his own right.

DARRYL STARBIRD: Are there some cars that I wouldn't chop? That's exactly right. It's always bothered me to see the guys chop the Mercurys as much as they do. A couple inch, or three-inch chop on a Mercury is not too bad, but even then, you really need to do the body also to get the thing back in proportions—which most guys don't do. And the radical, chopped top look is a fad. Especially with the younger, for lack of a better term, rat rod boys, they really like a chopped look on everything, not just the Mercs. And they certainly don't care about proportions.

Where else should you cut on the Merc? Right through the body. You section it. I've done two sections over the years. Back in the '50s I did the first section ever. I guess it was '57. And that's when Barris and everyone was doing all those chop jobs. Nobody had done a section job. What motivated me to do that? Well, again, it was strictly proportions. I saw all the top Mercs that Barris did, and everybody loved them. I am not saying I disliked them. I just didn't feel they were proportioned properly. When I first saw them in a magazine I thought the body needs to be smaller for the head. That's what inspired me. Just looking at the pictures and thinking about the proportions. In fact, that first Mercury that I sectioned I had first chopped with a guy named Jerry Titus, the KKOA [Kustom Kemps of America] president. He owned the car at the time.

Again, after we chopped it, for me it didn't look right. I finally talked Jerry into it; he actually drove it for almost a year before we sectioned it. I'll take that back. He sold the car and a friend Gary Carr owned it. I sectioned the car with Gary. Later, Barris had a model car made of it and put his name on the side of it and said it was a Barris Merc. That's a kick I still get out of it all the time; he'd say, "This is yours" and I'd tell him it's the best Mercury he ever built.

How'd I know where to section it? I had just had a hit and miss situation. I know that when you drop anything that you have to find the flattest possible point. When it drops, it drops equally down. I found the flattest part of the line so the bottom contour equaled the top line and would still come together when you chopped it. And Mercs are pretty round. I went right below the accent front wheel well opening and then to the rear wheel opening.

Did I have any reservations about making that first cut? Ha! Everything I've ever done has been a challenge; I've never hesitated. There's always a way to get there. You know, that question has been asked many times. I've chopped up brand new cars, like the '59. I bought a brand new '59 Buick from the dealership and took it to the shop to customize it. That was always my mindset, to do something to them, regardless of what people said or the possibility of not making a profit.

Have I ever had a car that I got into and thought, "No, this isn't going to work" and just put it away? Ummm … no. I had numerous cars I started without funds; at times finances were a problem. I can't honestly say that I ever started a car that I just felt I did the wrong thing and just give up on subsequently. I have, as I got older, slowed down. It takes me longer to create 'em. I've got that sport utility sitting over there for six or seven years. When I started on it, the sports utility thing just started. I jumped on it and then I got sidetracked with other jobs. And I kept getting sidetracked. It's not a real priority to me, as I've gotten older. In the earlier days, it was always get it and get it done. As soon as possible. Right here. And fourteen days later, it was done.

My grandson Cody has picked up this. He has a real good work ethic. He's learned a lot from his grandpa. Both of my sons were that way. They work hard. Too hard. Did Cody get the Starbird gene? He seems to have it, for some reason. Before he moved to California, him and his brother were in the shop. Cody was in college, and I said to him, "You've spent the last three summers in a row with me." And he said, "Grandpa, I've spent the last NINE summers in a row with you." I've had a lot of influence on him. Debbie, his mother was single until he was older. But, together, we kind of formed his character and work ethic. He can do the work of two people in the shop. It's a gift. And he's managed to capitalize on that gift. He'll just keep on developing it.

Darryl Starbird in his office at the National Rod and Custom Hall of Fame Museum in Afton, Oklahoma (June 2012).

My shop is behind the Hall of Fame build-

Darryl Starbird and his grandson Dakota Wentz during the chop of my '58 at the Hot Rod Hall of Fame in Afton, Oklahoma (June 2012).

ing. I kinda just like to just go there to the shop and just hide. But often I'll be asked to come back to the museum building to meet some of the visitors and sign stuff. That's just part of the territory. I enjoy it some days; if I didn't I'd do something else. I've had the luxury of doing what I want all along. But luck has played a part.

Like the first Monogram car model deal. I had this Forcasta, which was my second bubble top car and it was on the cover of *Car Craft Magazine*, which of course was the big custom car magazine in them days. And Monogram Model, at that time was just beginning. They was thinking about getting into the hot rod custom car model world. And doing models of it. And their first effort, and I never did even ask them why, was a V8. An actual working V8 model. But their next thought was to do a ⅛ scale hot rod. And then custom cars from then on. Well, it just so happens that when they did that first ⅛ scale, they ran their first ad in a hot rod magazine, which was *Car Craft*. And it happened to be the magazine I was on the cover of. When the president and the vice president saw the magazine with their ad they said, "Boy, that car on the cover is unbelievable. That's the man that we want to be our design consultant for our models." And out of the blue in my shop in Wichita, Kansas, Jack Best, the president of Monogram calls me up and says, "Darryl, love your work. I want to come down and see you in a couple weeks and talk to you about doing some stuff for us."

At that time, I was in a little a six-car garage. Didn't even have an office. I haven't progressed much. They did come down, and I did equip a little office to meet them. Well, they were enough impressed with the cars that we were working on, as well as that car.

The August 1960 *Car Craft Magazine* cover featuring Darryl Starbird's first bubble top car, the Predicta.

I am sure they wanted to inspect the shop; they were a little concerned that I was small. Anyway, long story short, I did sign a three-year contract with them at that point. They renewed it, so I had six-year tenure with them. I was their design consultant with a salary. And later I got a percentage of the kits that was sold. That lasted 15 years maybe.

Then Mattel purchased Monogram. They made the Little Coffin into a Hot Wheels diecast toy. That got popular right away and that kind of changed the whole face of the model car thing. Then model cars disappeared when computers came in. And video games and all that. And kids quit gluing things together, unfortunately. And the plastic car thing just went away, and then Hot Wheels kind of picked up. The diecast cars. In fact, I was just inducted into the Diecast Hall of Fame. Just for my stuff made in diecast. Anyway, that's how that went down.

Did my engineering background help with the plastic model design? Well, I understood how it worked, because I had worked for Boeing airplanes. I'd worked in numerous departments; I'd had a lot of variety and exposure with all different types of engineering and construction. That was a good opportunity, my first job out of high school. My dad worked at Boeing in layout and tooling. He'd spent his whole life there, 40 years. Anyway, that's where he wanted me to be. And, of course, like a good son, I wanted to be an aeronautical engineer. Although my first love, even at that point, was really cars. In high school I built me a '41 Ford. There were only two custom cars in the whole high school; I had one of them.

But anyway, I went on to university. I still worked third shift, which was 1:30 in the morning till 8. We worked 6½ hours and got paid for 8. And there were several of us—about a dozen—that were all going to be engineers. And Boeing kind of took us under their wing. And consequently, always kept us, even when they would discontinue the third shift. They'd send us to another department, so we could continue our studies. That went on for almost four years. At that point I was just done with being an indoor engineer. I had spent a hundred hours designing and building a valve and my supervisor just threw it in the trash. That was crushing. I knew right then that as long as I was going to be an engineer that I'd always be working for somebody else and they would have control over things. I just didn't want that. I wanted to be self-employed, to be my own person and to be creative in a way that I wanted to be. To take the hits however they came. And it wasn't easy in the beginning. I had no name and I was in the middle of nowhere.

Where did the idea for the bubble tops come from? I had the design idea first. The first thought I had was when I worked at Boeing, as a matter of fact, was I wanted to build a futuristic car. It's what I dreamt about as I sat there as a draftsman. My first futuristic car; it always had a bubble top. Or a glass top. And my only path to get there—that I could see—was the '53 Studebaker Starlight Coupe. That had wrap-around windows around the back of them. And I was going to take half that top—the back half—and make the front. If you brought two of those together, it was almost a bubble top. That was my first thought for a bubble top. But then I went, "Well, with that top, was it was going to affect the proportions?" There was a little sketch I did, and the squared up look just did not fit what I had in mind for the car. Anyway, I had to have a bubble top.

My next endeavor was to try to figure out how in the world I would ever build a bubble top. I accidentally got on to a guy who had a sign shop. He made plastic signs. I was talking to him one day when I was having him make a sign for one of my cars. I said something to him like, "I'd sure like to make a bubble top." He said, "Well, why not?" I asked him how'd he do that. And he said, "Well, I used to work for Bell Helicopter and they did domes in a bowl form." He explained the theory of it all to me, and I kind of just improvised the rest. And we built our first oven big enough to hang a piece of ¼ inch sheet of Lucite plastic. And it was a big sheet—even the Predicta was huge—it was 6 by 8 feet. And I had to figure out how to heat that to the proper temperature.

We built our first oven with his help and guidance. I built the first two of four. It was tricky; the temperature has to be very evenly heated. That was the biggest problem. The bottom half could burn. Anyway, theoretically if you took a sheet of plywood, cut a hole in it and take a piece of plastic and blow it up to half the height and width of the hole, you have a perfect half circle. So that's the theory that you gotta go from but you have to alter it to get the shape you want. Otherwise it's a perfect half ball, which Roth never got past. You see his cars; they were just a half globe, different diameter circles. As you alter the shape of the hole you get a different shape, you begin to alter the shape of the bubble. The thing that makes it take a shape is the restrictor opening in the middle. Your restrictor has to match the profile that you want. You throw that big piece of hot plastic on the big table and the table has a hole in the middle of it for the air valve. You just throw air pressure to it. And you kind of inflate it to push the center of the hot plastic sheet through the restrictor. And whenever it gets the height you want it you just cut off the air pressure. And then it just cools in its shape. You can't go past half the height and the width or it begins to blow out instead of up.

The first time we tried it, it didn't work out. We never could get the evenness of it; that was our biggest problem. The plastic would not heat evenly, and it would blow more on the one side and miss on the other. This thing took a lot of air pressure. The second time? I don't remember exactly, but again we didn't make it. We even ended up buying another sheet of plastic because when you reheat it you begin to lose your optics. We were losing optics on the second try. On the third try, it almost got there, but one side was just a little bit higher than the other side. I didn't like that so we heated it up—it has a lot of memory—and it went back to a flat sheet. But by this time, of course, the National Roadster Show was in Oakland, California. This was my deadline, to be at that show.

That was my plan. I had the car all done except I didn't have the bubble to go on top of it. I had to get it there on Tuesday to get the car set up. So I left without the bubble top on the car in the trailer. It's about a three-day trip to get there on old Route 66. And anyway, I got there with no top. But by the meantime, the guy had tried a fourth try and it come up perfect. Without me being there. And he shipped it to me. By the time I got there, my top was there. But the problem is that you still got to trim it and everything. And build a ring and all that. At its very first show, the Grand National Show, the car was there and the top was sitting next to it. It was missing the top. Which wasn't all that cool.

But my biggest problem was that *Car Craft Magazine* had also scheduled the car to be on the cover. And they was waiting to shoot it. Well, I had at least a couple days and Bill Cushenbery had a shop in Monterey, which is halfway between LA and San Francisco. So I went down there with the car to his shop, and I got a router and we cut the flange off the bottom. We put it on the car, and it looked like a bubble top, although it wasn't finished and I hadn't built the ring around it to fasten it down. I went down to LA with it and the editor of the magazine at the time, Dick Day, and for the cover shot he was sitting in the passenger's seat and I was sitting in the driver's seat. And he was holding his side of the top and I was holding the other. Nobody who bought the magazine knew we really didn't have the top on the car. And we did shoot it with the top off. The whole magazine article was with the top off. Of course, I took it home and finished it all off. That's the story of the first one.

I did not keep that car. I built it in '59 and kept it till '63, when I sold it to Monogram. Then about three years later—about '66 or '67—they gave the car away in a model car contest. To the guy that built the best model. The crazy part about that story is a guy named Darrell Zip won the Predicta. He was a designer for Revell, which is a competing model company. The folks at Monogram didn't realize that when he won it. Then, the guy that won it traded it for a high-dollar chopper motorcycle. The car set in that motorcycle showroom for two or three years. I heard about it. I was out there producing a show, and the guy called me up and said, "Darryl, I got your Predicta, and I want to sell it." I said, "Really?" And I got the whole story. Sight unseen I bought the car. When I got it I was disappointed. They'd painted it silver metal flake, they'd put a hole in the hood and put a big blower on it. They'd turned it into a Mickey Mouse looking thing which I took home and stored. That was probably '73, maybe, when I bought it back. I've had it ever since; I've restored it twice since then.

What did I feel when I got it back home and it was messed up and the vision had been carved on by somebody else? You know, it was great to get it back. For a year or two I had no idea where it was. I mean, I kind of lost track of it; nobody seemed to know much about it. Then all of a sudden I got that call. Would I say that's my signature car? Sure. It was one of the cars that put me on the map nationally. It changed the direction of custom cars. And for a few years everybody was doing bubble tops and experimental cars. Even Barris, Winfield, Jeffries; everybody jumped in for the next four or five years doing bubble top cars. It destroyed the iconic early '50s custom. Up to that point, everything was the image of the '50s Mercury. The Predicta changed the whole car scene; you had to do more than work on Mercurys and '40 Fords. Folks began to work on late model cars to build futuristic cars. And even the model car companies helped a lot; because they wanted futuristic models because they sold well. And experimental types. Detroit was real big into prototypes at that time. So it all just kind of happened, and I just happened to be in when the timing was good.

Do I chuckle about that? I do because, unfortunately, commercially my cars, other than the model cars, haven't made the same impact. Everybody now wants a '50s type custom. My trend has kind of went away. So commercially now, especially with Barris in particular, everybody has been able to capitalize on his style. Now today you can talk to almost anybody in the industry or anywhere and when you say "custom car" they think of Barris cars; they don't think of mine. Mine is later on, futuristic cars. But commercially, Barris has done the best, in terms of cars. Now as a businessman, I've done just as well as he has. But as far as a "caught-on" statement, mine have faded.

For some reason, when you get right down to it, there's not too many people who have the skill and the nature to do the type of work it takes to build a futuristic car. There's only a couple today who know how to do that. There was Boyd. And, of course, Gene Winfield. He's near retired like myself. Chip Foose; he has the ability to build any car. But I'd have to work to count them on one hand. Others? If they have the ability, they never explored that ability. There's gotta be some more out there that probably just haven't had the opportunity.

But the metal crafts part? You know there's been some guys that have tried fiberglass.

And they've built some pretty decent cars. Of course, it's a different method of construction altogether. But to be a true metal man building brand new cars, there's not too many out there. Can my grandson Cody do it? He has the ability, he can do anything. On the car I am working on, he did the inside door panels and that took forever. And he did the top, all day long of hard work. I'd have done it in two hours. Is that practice, experience, and the good eye developed over the years? Probably all of the above. But he's getting on it. If you saw when he brought his black truck here three or four years ago now—well, I had to go over every bit of it. I mean he'd dress, and dress, and dress all day long and it'd still be wavy. I'd go back to touch it up for him. So even with that truck, when he was done with it, I'd just make him keep going over and over it. Until he finally would get it straight.

Getting 'em straight is an art. The skill, or whatever you want to call it, just takes time. The sad part about it, and I am not mentioning any names, is that some people just don't have the eye or the feel. And the hands. I can feel anything you know? Cody's got where he can now. But that summer he couldn't; he had no idea what he should be feeling for. It's gotta develop; your hands become very sensitive, in feeling it. And, of course, your eye too.

I've got a good example of that. That door over there [pointing to a door on the Shark], I've redone it now twice. I've got a ripple in the middle of that door. A small ripple that you don't see unless you look at it just right. I am on my third try. And there's no way in the world my hands can pick that up. I mean it feels perfect; but it's there. I've primed it, blocked it, and put the black paint on it. The worse thing is the other three doors, I had them done right the first time. They are perfect. So even then, it's tough to get them perfect. And black is the worst.

Just recently House of Kolor has come out with what they call Shimmering hues—and their black is black, just the way I want it. They even say it is blacker than black. The thing you don't do is add tint to clear. If you just put clear over black it gets milky. It kills the black to some degree. It will look real shiny and deep but it has kind of a cloudy look, because the clear gives it a little cloudiness. So you tint it a little bit black. And the last coat in clear with a little bit of black.

What's a hot rod? What do I build? Starbird cars. I guess I've been an automotive stylist most of my life. I put a lot of style in them, not only body-wise, but frames and things like brackets. I do all that in a different way than anybody else. You'll see that if you get under one of my cars and look. Most of my cars are bubble top cars or prototypes. Experimental cars. Dream cars. Or concept studies of design. My other thing is, well, I like to tweak 'em. Like the '59 Buick or a T-Bird. I would pick, over the years, what I felt Detroit did the best job with and I like to tweak them to give it the Starbird touch. I've got a newer Cadillac XLR that I'd like to tweak, but the biggest problem is the composite body. It's not metal; that bothers me. I am not really crazy about cutting up a composite bodies because you don't have the "bottom" necessary to somehow get it back together. Composites are a layered and pressurized material. You can't work on it like with fiberglass or metal. It's a totally different thing.

What would I do to it if I had the technology? Well, I hadn't bought a new car since '97. I hadn't liked anything that's come out. I bought the Lincoln Mark VIII in '97. I thought it was still a good looking car. But from that point on, I didn't feel that anybody

Darryl Starbird working on his 1957 Cadillac Brougham custom, the "Shark" (June 2012, Afton, Oklahoma).

built a car that I'd want to buy new. Not that I did not have the money; I just didn't have a desire to own one. I said that whatever company comes up with something that I really like and I feel they did a really good job designing, I'll buy it. And when I saw that Cadillac XLR, well, they've done just as well designing that car as I could do. In my own mind that is. I am saying that car is as well-designed as I can do. I feel no need to redesign it.

I met Wayne Cherry, one of the major designers of Cadillac in Detroit. He is a tip-top designer. Now with most Cadillac car designs he is restricted. Or he never gets to express himself; but on that car he did. Commercially, it targets the older driver. They sold just a few, for a number of reasons. It's competing in the market with Lamborghini and Ferrari. But Cadillac didn't step up mechanically to compete with those cars; they didn't have the money. Well, they have the money, but they are not going to get $250,000 for a car. Not in this country, for an American car. And that's sad.

I don't know what I'd do to my XLR. I am sure I could do things to it. Usually that's the first thing I think about when I see a car. I want to change something. It's kind of a twist, it's different; they kind of blocked me on this one.

You ask me what is my message to future hot rodders, what is a message I'd put in a bottle for them to read 50 years from now? That's a good question. First of all, 50 years from now, or 20 years from now, it's probably going to be a totally different car scene. We'll be seeing cars in a different way. I suppose the American Dream, which includes the automobile, is going to be changed. In fact, the museum here is dedicated to keeping the history alive. Which, of course, will fade if you don't tell future generations what it was and what to do to revive that if they can make it happen. It's probably going away now. I really don't know what to say that could mean anything. I mean I could say never give up building special cars. But if there is no cars around? It's kind of like the people who used to build horse buggies; what would you do with a buggy now? But I do honestly feel cars close a chapter. They will be driving themselves with remote control that will stop and start automatically; moving down the highway we'll still have individual transportation always. Well, for a long time.

What's going to happen going forward? You know, everybody says that customs or cars are going to go completely away. But on the other hand, you think there's still people that are restoring Model T's. So, 100 years from now they will still be restoring '50s cars. And of course they'll be '90s and 2000s cars then too. I think that sport, or that hobby, will always be there, but maybe not to the same degree. They will build cars with some of this modern technology. But car guys will stop at a particular year, and only restore behind that. Kind of like what has happened now; everyone is doing up to about '70. From then on they could care less about it. You can't change the length of the car from that point on. But how many more years will they be doing the '70s cars? I am sure it's a hundred more years. Just like it's been almost a hundred years since they did the early 1900 cars.

So I think our sport is sound. It's a niche market. Will it be a smaller niche market in the future? Exactly. Maybe the technology will help. Like a laser tool to find the ripples. There will be shortcuts. But proportion, the good eye, the fingertip touch, and the ability to make things "pop"—technology cannot replace that. One of my pet peeves, so to speak, is I have people come in here every day almost and say, "Boy, that Chip Foose is an unbelievable designer." Well, I don't say this, but in my own mind I think, "Yes, Chip is an amazing designer but he is a proportion man. Period." He went to an art school in LA; all they teach is proportions. And that's exactly what he does to a car; he puts it in its proper proportions. An eighth inch out here, seven-eighths inch out here, two inch out here, and he ends up with a car which is in perfect proportions. Which is great; don't misunderstand me. But that's not design.

Design to me is something original: A new thought. To be a designer you gotta have original thoughts. And even in Detroit, the designers over the years, most of them just copy what's been going on. Every now and then somebody will come along who is a real designer, with original thoughts. I haven't seen it too much in the custom car world. Especially now. Everybody's just taking a '70s and getting done, it looks like a '70s. Or a '60s or '50s or '40s. I like to see an original design. A designer, in my opinion, isn't someone who just fixes 'em up or tweaks them. And to me, this is where most of the guys are at.

You ask me if Cody could become a designer? I have a little problem with Cody. And I have tried to kind of guide him in terms of what direction. I keep telling him you got to think original ideas. You know that blue '55 truck? Cody is caught between an original idea and the '50s. He's still buried his thoughts in the '50s. And he wants to take those thoughts and just tweak 'em. Make them fit the contemporary. If you read the article in *Custom Classic Trucks*, that's what they say he does.[1] But I have yet to see him step up to the plate on a car.

How does somebody get out of the '50s custom stuff to take the next step? Cody has all the technical tools and the true design eye; there's no question about that. He just

Darryl Starbird's "Cristina Mark IX," a custom-built 1941 Lincoln Continental, completed to celebrate his 50th anniversary in the custom car building business (July 2016, photograph provided by Donna Starbird).

has not got the original thoughts yet. He needs to develop them. Maybe he can't develop them. Maybe that's a gift that only a few people have. Maybe he does not want to be seen as mimicking my career path? That's possible. He is set in his mind that he wants to be recognized for his own work. And he should be. He'll have to struggle with that.

I am not saying it's bad to be that way. Like Foose; he does great work. Amazing stuff. But I am just saying there is, in my opinion, a difference between a designer and a person who just tweaks somebody else's design and makes it a little different and better. Just like this '56 Lincoln Continental Mark II that I am working on. I didn't design it, all I did was take it and tweak it. It's the original design, I just took it and updated it here and there. All I am saying is that a real true designer does more.

If you really look at my book you can see my progression.[2] You can see where I started, just like Cody. You can see how my designs progressed over the years, you know. Things changed, and I went in different directions. I've had two different directions. I've had the commercial direction; I always felt a need to do what people liked at that time. I had a feel for that. Not '50s customs or '49s, but for really unique cars that people would want to come to shows to see. That was my driving force at that time in history. And then the other direction was to just be more creative. And I've been more creative in the last 20 years because I've had the freedom to do that. To do what I want, without having the commercial aspect pounding on me all the time.

But still, I have people tour the museum here and walk out all the time saying, "Boy, Cecil the Diesel, that's the best thing you ever did!" And I think it's the worst thing I ever did. But for the time it was what was going on right then in the '70s. Everybody was building that kind of stuff. So I had to do something to show them that I could do that if I really wanted to. That's when I came up with Cecil the Diesel. But it is amazing how many people still think that's the best thing I ever did. It's almost a slap in the face. If I can't do nothing better than that I'd better pack it in. But it's cute. A theme car, so to speak. But today, the general public doesn't have much eye for design, that's for sure. A car is just a box to get to their next destination; but maybe that's all they need today.

11

Brylen Brajkovich (Jonestown, PA)

Brylen Brajkovich owns a rod and custom shop in Johnstown, Pennsylvania. I met him at the 2009 Carlisle All Truck Nationals, which I covered for Custom Classic Trucks magazine. He had three vehicles in the Invitational Display building's Rat Rod vs. Traditional Custom exhibit. One of his vehicles was a mild custom '53 F1 shop truck. The other two were rat rods right out of a Mad Max movie. But all three were well-engineered and nicely finished. One was the Puppycrusher, a '27 Ford built for a Cruella De Vil (of 101 Dalmatians fame) movie. One reviewer called the rod a "twisted and diabolical mashup." It won an Extreme Award at the 2008 Detroit Autorama.

BRYLEN BRAJKOVICH: I grew up on a farm; we were dairy farmers and chicken farmers. It was rough; you get up at four in the morning to milk and then bale hay. My dad always looked at cars and trucks as just modes of transport. He wasn't really worried about beat-up fenders. But I loved cars. When I was a kid my mom got us those Little Golden Books. She bought a stack of them. You know what I used them for? I stacked them up to reach my dad's models. What, you supposed to read those things? My mom finally gave up; she just gave up. I'd fool with Matchbox cars.

Matchboxes, you know with the square box? I used to keep the boxes; after I played with my cars, I took care to put them back in the boxes. Weird. Then I cut the roof of one and it looked better as a roadster. The rest is history. I blame Ed Roth for most of it. He was at the top of his game then. I remember his Beatnik Bandit and stuff like that. I wanted the models, I got the models, and I started putting them together. I'd see those cars with a bubble top and I'd think, "Wow, what planet is that Starbird guy from?" They would blow my mind. So my work was influenced by '50s customs.

I was a normal guy through high school, had shop class, and did woodwork. But my grandpa was a woodsmith. He did fine furniture. And he always used to tell me you can always re-weld metal, but once you screw up a wood piece, you are done. So I took that knowledge and I thought to myself, "Hey, let's try building cars." I built my first car when I was 16. At the time, it was a very unpopular car: a Mustang II. I got an education building that car. Anyway, went to a VoTech and took welding classes. And then right out of high school, my dad said, "What are you going to do with your life, 'cause you are not staying here for the rest of your life." He said, "I'll sign for you; you are going to do something. I'll get you started."

So I bought a big truck, and I started hauling cars for Detroit. And me being an older car guy, liking hotrods and muscle cars, the mid '80s stuff was not doing anything for me. There was a big void there; cars were like blaaaaah. You know, like Escorts and Sables. I was screaming for something else. I sold my trailer and kept my truck and started hauling antique and fancy cars. I went to Hershey and Pebble Beach and started hauling for Robert Pass. That's what got me out to the West Coast.

I started hauling cars, met a lot of people. Got into the movie stuff. I hauled for Stephen King. Maybe I set some kind of foundation I did not even realize. I was on movie sets when I was hauling cars. Because that's what you do; it's a hurry up and wait thing. I did that for some years. And being into cars, I went to a lot of car shows. So I met a lot of people, like Pete Chapouris. Eventually I blew my truck engine and got a job in a wrecking yard. And a couple of movie stars used to come in there—they used to film there. I met Sly Stallone there, a real cool guy. He says, "I got a car I am working on. Got a Corvette; it won't start." It was a '63 Stingray. Beautiful car, simple starter issue. I fixed it, and next thing you know, he's telling people, "Hey, this guy can fix cars." So I started repairing cars.

One day Pete Chapouris called me and said, "I hear you are out of work. Do you want a job?" A year earlier, I met Pete at a car show. I was hauling a '32 Ford. He said, "That's a cool little hot rod you got there." I explained I was just hauling the car." He said, "You got a business card?" I wrote my name and number on a napkin, folded it and gave it to him. I couldn't afford a business card. How pathetic is that? I could barely pour gas into my truck to get here to the other end of the block. Anyway, so I handed it to him. He kept my number. And, a year later, he called me. I worked for him for two years. And you know, I didn't tell them that I could weld for the first two months. Because I was afraid, because I thought, "I can't stand up to these guys."

Eventually, I got totally burned out and got out of the car business for two years. I moved back East and started working on fire truck cabs for American Lafrance. One day Ken "Posies" Fenical, the owner of Posies Rods and Customs in Hummelstown, Pennsylvania, comes in and sees me TIG welding. He says, "Good job. Any chance you want to come work for me?" My foreman gave me a bad time for taking five extra minutes on my break. My middle finger went up, and I said, "Fuck you. I am done." I swear to God, it was time. And it felt good to pull out. But I am a real serious guy, so I try to give them notice, and they say, "Get out." Ken says, "Screw that, you are working for me."

It definitely was a turning point. That was four years ago. I knew about Posies, because I had one of his car pictures hanging on the wall when I was a kid. I used to look at that car and dream of driving it down the highway. Or think, "I'm going to build a car like that someday." That was my little dream. And there he was, offering me a job. I was with Posies for three and a half years. I learned a lot from him, I learned all the stuff that you don't do and all the stuff that you do. Posies is a good businessman. He is an awesome craftsman. Vision? Can't praise him enough. Personality-wise? We just never really got on. But I respect him.

Anyway, so I left him. And I swore the next time he'd see me it would be at the

Detroit Autorama and I'd win something there. I'd build a car and win one of the big awards. Now I know I'm not going to win a Ridler, nothing big like that, but I figure one of the traditional hot rod awards in the basement. So I left to start my business.

And now, here we are. My shop started as a 2,000-square-foot shop on industrial land, an acre that my grandfather owned. It was a very valuable property, just off an I-81 exit in southern Pennsylvania. It is a great location. Red Lobster offered to buy our place twice for a whole lot of money. But my grandfather once told me, "Don't sell out. The time to sell is when you can't do it. If you can do it, keep going; don't sell out just to be a bum." He had a lot of little mottos, little one-liners. But that was one that stuck in my head. And we've been there, now it's a 4,500-square-foot shop with a 2,500-square-foot showroom. I added a new machine shop in the last four years, hired on two other guys, even with the struggling economy.

I promised my wife a vacation, so we went to Vegas. We'd saved money for the trip. I was playing the quarter slots and this guy sits next to me. I was wearing a hot rod shirt, so the guy says, "Hey, you have a hot rod shirt." He pulls over his chair, next thing you know he has his wallet out and he says, "Look at my cars." No kid photos: cars. He talked about his '32 Ford and told me he worked for an LA record company. I'm thinking, "This

Interior of Brylen Brajkovich's shop. The orange 1972 Camaro is undergoing a full pro-touring build, in foreground is a blue 1930 roadster pickup, and at right is a 1928 sport coupe with blown small block Chevy (May 2017, Jonestown, Pennsylvania. Photograph by Brylen Brajkovich).

guy is for real. He might not be blowing smoke up my ass." I told him about my shop, and gave him my card. But I really didn't think much about it.

Next thing you know, I am building three cars for a Cruella De Vil movie. I built them as rollers, sister copies, in just three months. Shipped them out in primer; no steering column, no pedals. Just rollers with a lot of parts on them.

I got to keep one of them. I called it the Puppycrusher and it hit it in Detroit. I had the car on the trailer, I had it covered with a sheet, over some wash poles taped on the fenders to distort the features. It looked like a box, a big tissue box with wire wheels. I pull in, and next thing I had Chip Foose coming up and saying, "Hi, Bry. How you doing? I remember you from LA. What's under the sheet?" And I said, "Tomorrow I am going to pull the cover and you are going to see."

It was cool. Here you had Chip Foose, saying, "Let me look under your tarp." And what's under the cover wins the Extreme Award. Wins a trophy, the whole deal. It was a huge thing; I couldn't even lift it. I got a check, airtime, and web coverage. It made a big splash. I have a picture of Gary Meadors pin-striping the front. He's sitting on the cowl. I said, "Gary I can't afford you." He said, "Yes, you can. I want to run lines on the car." I said, "I'll take that."

I call the Puppycrusher a coach-built car. It's a hand-fabricated wonder, made in three months. I built the car and it dominated my whole life. For three months, my wife did not see me and my daughter did not see me; that bothered me more than anything. I got angry at it. Angry at a goddamn car. For Christ's sakes; pissing me off, fucking my life up. I just wanted it finished. And I had nightmares, the damn thing chasing me. It was ridiculous. I had a guy at Detroit come up to me and ask, "You built this?" I told him I did. Then he said, "You wake up at night screaming, don't you?" That's what he said to me. It's true, that car consumed me. I lived at the garage; a lot of times when I worked on it my wife brought me dinner. But that's the business.

On the business end of things, my shop builds more trucks than cars every year. We just finished a '27 Chrysler Reo pickup for a local attorney. Who would have thought, you know? We put an injected small block Ford in it with a six-speed, and the guy's tooting around town in it, getting stares and gawks that his BMW can't do. And, you know, he called me up on his cell phone, cruising his Mercedes, and he says, "You know what? My old hot rod gets more looks than I've ever had in any vehicle." And he is a very European-influenced guy; classy. Drives a nice Ferrari and has all the high-end cars. But this old '27 truck we put together for him? He loves it.

You have to look at building cars as a business and a lifestyle. I live the lifestyle. I am not one of those guys that jumps in the Mercedes and leaves. No, I jump in my 30-year-old shop truck and leave. And that's what I do. It's a tough business. Things did not come easy. You can love a car and put way more money in a car than it is worth. On the business side, I have to grab myself and say, "Whoa, whoa, whoa. You are getting too far ahead of yourself. You are putting $50,000 into the car that's worth $15,000 at best." It took me a while to sort this out.

How do I balance things out? Well, I've got some rounded customers, but also some on a tight budget. It's a mix. I need to keep things flowing. I see two empty bays, I get nervous. My guys need to feed their families. So if I get a customer that needs parts for his hot rod, there's potential. And you always have to put something on the shelf when

Brylen Brajkovich with his '27 Ford, the "Puppycrusher," at the September 2016 AACA Museum Show in Hershey, Pennsylvania (photograph provided by Brylen Brajkovich).

you are a businessman. You still have to make money. If I can do a set of wheels for his truck with polished centers with spikes, I'll do them. So I sell parts. But what I am also going to do is take this guy into the shop where I am building hot rods. He might say, "Wow, look at those bumpers!" And I say, "We can build them for you." If a guy can't afford a $20,000 or $40,000 build, he may be able to afford a $200 bumper. That might lead to something else. You sell parts so those two empty bays are offset.

So parts are another revenue stream. I see a lot of potential there because—let's face it—there are a lot of guys in home garages across the country who build hot rods and trucks. They want to do it themselves and I praise them for that. I can help them with what they need. I just purchased a new machine, so I can fabricate motor mounts. We're moving into that. So, yeah I'd love to have a catalog in ten years, say seventy pages.

Right now I am building a '53 cab over for a customer. And I'll give you a little hint on the street rod building business: when I build something, the process itself attracts other customers. If I see six months down the road that things might be drying up, I get very nervous. Because I am not making any money then, you understand. I have four guys with me; three employees and we got the broom guy.

But size-wise, I want to stay about where I am at. I got three guys, six bays, a small showroom with some parts, a car on display, and a small register. It's a mom and pa thing for me; it's hard to keep three guys in line. Imagine 30. Because you are a shop builder, you become their therapist, their buddy, and a lot of things. I don't know if I could run it all. I am only one guy, so I can't say, "Hey, go to the office and tell it to Margaret." I can't pawn them off; it's not going to happen. So I got to handle it myself. And my wife, she can only handle so much; right now she handles the books and runs the show. And what the heck, she works a regular nine-to-five. Amazing woman. God, I mean, who would ever stick with somebody who says, "And by the way you are going to work every day, and then do this?" How many women would say, "Oh yeah, I am going to jump at that opportunity?"

I've thought about getting bigger. I struggle with that notion. But I also cringe at the thought. When I built Puppycrusher, at the back of my mind was the thought that this could launch us into space; here we go, 40 bays, 60 guys. I could be the next Speedy Bill. But I kept that in check. Maybe Speedy Bill on a lot smaller scale. Honestly, I really don't want to be the one who has 50 guys working for him.

You ask me to help you understand builder/customer relationships. Well, there are certain situations where you have great customers. Tommy Gunns, the former world heavyweight Underground Fighter title holder and actor. Tommy is an example of a great customer. He's a fighter, but he's mechanically inclined and really into cars. At the other end of the spectrum, there's someone who had a particular car in the past but is now retired and wants that car again. There's 30 years he missed, sitting behind a desk. So he comes to me and he says, "Brylen, I am going to trust you with a lot of my money, and I worked hard for it. So this is what I want...." We started talking, and at first, maybe we don't get along. Not at all. You know, Jersey guy, bad attitude, aggressive, and so forth. But I have attitude too, so it works out well. Because if you are a timid guy, you are talking to him and you get nowhere. It's all business: In the end, you are going to spend that guy's money. So you get the relationship worked out.

Sometimes I have to ask a client for more money when I run out. How do you ask a guy for another ten grand? It's about taking the car to the next level. I say, "You wanted this level." And the guy says, "I don't understand levels," that kind of stuff. I say, "You want a car that's gonna win at Carlisle, or go to a SEMA Show in the Flowmaster booth? To get that, you have to spend more." It's not all about labor, it's mostly about parts.

Don't get me wrong; labor is not free. But I tell him, "Trust me when I tell you, in the end it's definitely worth it, it's going to take it to that level. So, OK you are going to spend a lot of money. I am going to do this and that. I am going to call you every day, and you are going to call me every day. Sometimes we are going to be butting heads, but we'll get this done right." So the relationship works. And in this case, we build his car, a '59 Chevy. It was invited to Carlisle, and it got a front row invite to World of Wheels in Detroit. And it will be in the Flowmaster Booth next year. As promised, I delivered.

Almost half a high-end build is labor. Almost. You spend more on parts because you do a lot of custom-machined parts. You want a 572 Chevy big block? That's $10,500, not even getting out of the crate. That's a credit card swipe, money gone and it's on the way.

The easiest guy to build a car for is the one who walks into the shop and says, "I want you to build a car for me for a $100K budget." We sit down and work out the details. This kind of guy has done his homework, he's been there, and knows what it costs. So does the builder. You have to watch out for a builder whose eyes get real big at the mention of $100K. You don't want to get $50K into something and have the builder say, "Well, I got $30K from him and it can sit in a storage unit for the rest of its life. I don't care."

Reputation is everything in this business. I want a car to represent me. I want the owner to be happy. And I want cars I build to be a good investment. I want to be able to flip the car if I have to, to sell or trade it, whatever. But I want it to hold value. I have people from all walks of life come into my shop: police officers, lawyers from New York City, and professional fighters. I recently built a roadster for a school teacher. Whether it's a $100K build or a $30K build, you get it done.

Let's get more budget minded; more working class like me. You have a budget of $20K or $30K, and you want to build something that doesn't break the bank. I can build a car for a budget, and still produce a cool car. But this is the hardest car to build for the money. Think about it. Here's how it is. I sit at my desk before I start a build, sit down with the prospective customer, a guy just coming in. He doesn't know me, I don't know him. I am sniffing his butt as much as he's sniffing mine. From the business end of it, after the first five minutes I know whether it is going to happen or not.

It can get complicated. A guy comes in and says, "I've got $30K." He shows me a picture and says, "I want to build this truck like this." Now, I am not going to build that truck for $30K. I tell him that if you do some work yourself, maybe if you chop and section it, if you do the S-10 chassis, and then bring the cab up, then maybe. But you are looking at a parts bill of $4,000 for a crate engine, $1,500 for the tranny, whatever ... it all depends. The bodywork just to get things straight? Custom paint? Good God, candy paint alone is $3,000 just in the can.

I have a car in my shop, right now, a '63 Chevy; he's a younger guy, he's got money, but he only wanted to put $50K into the car. In your mind you are thinking, that should be a pretty doable thing. Not really. New quarter panels go for $1,200 apiece; he needed both panels. Long story short, when you start breaking the numbers down, his parts bill is something like $37K. That leaves me $13K for labor. Right there that 50–60 percent for labor just breaks down. Now I am losing, see what I am saying? Now I am going to have to put my business hat on. I might do the paint job cheaper? But that final product is my product. That's going to go out the door and people are going to know that it has my shop's name on it.

So now I've got to make some choices. I ask myself, "Is this car going to take me somewhere?" And if I think that it will, I come to the owner and say, "Tell you what. I am giving you a good deal, and that's not going to change. But as far as labor goes, I am taking a real hit. The parts cost $37K. The paint, if you decide to go candy, that's more money; you buy it. OK, we understand that I am taking a hit on the labor, so I need the car for six months after it is finished. And the client says, "What are you talking about? I am putting $50K in that car and it's not my car yet?"

At this point, they might turn turd-faced on you and want to strangle you. And I say, "No, I need to get my money out of the car." And the client says, "Wait a minute, you are not selling my car!" I say, "No, you are not understanding me. I need to show that car, I need to expose it. I don't want to motor it down the street to buy iced tea; I want to show that car. I want it at the big shows. I'll take full responsibility for the car; that's part of my deal." Most of my clients understand this. This is what I need, you know what I am saying? If you are a smaller shop like mine, you gotta have that because you don't have many cars going out the door, and every one of those cars better be a hit. Every car is a business card. I've learned to build well-planned cars. You need to do that.

I've held the reins back. I do a few cars every year. I go to a lot of car shows. But I still can't walk by a car or truck without seeing things I'd like to change. I just now walked by a truck that I could change four things on and take to a new level. I can't help it. Two weeks; change four things. I make the changes and you'd think it was a different truck. And it would be a whole new creature. Of course, it may not be what the owner wants. But I know I could take that truck somewhere; I could win Detroit with that truck. Give me two weeks, not much money, change wheels, grille, injection, and that kind of stuff. Every time I walk by something, I can't help it.

With a hot rod car or truck you are riding down the road in something different. It's about how you feel when you see somebody pulling alongside to snap a picture. Or when you see a hot rod or custom truck going the other way and you wave to each other. It's camaraderie, it's a family. I've stopped along the road for breakdowns, one time for an older gentleman with a '40 Chevy. This was about four years ago. I pulled over and he said, "You know what? Nobody would stop." He'd had a blowout, and he'd tore up the fenders because it was really low. We jacked it up, I gave him my spare, also GM pattern, and we got him going. And that was it. But no one else would stop. And he was not a young guy. A car guy will stop to help out. You know why? Because that guy may be driving by when I am broke down. That's one of the reasons. But also because you are family.

What is my definition of a hot rod? A hot rod is supposed to be a home-built car. Because that's the way it started back in the '40s when guys came back from the war. Fighter pilots had this adrenalin—you know—from shooting planes down. And they come home now and are just sitting in a chair? They are driving a box; it's driving them nuts. They'd rather be dead than doing that. So they get a '32 Ford and blow the fenders off, open the headers, paint it flat black paint, or red. Whatever looks cool. It's chopped low; the guy wants to stare at them injectors as he is going down the road. It makes him happy. I know it does; it makes me happy.

My definition of a hot rod is anything that fits your own personality without breaking your budget, OK? Because a hot rod is supposed to be fun; if you have a million dollars in the car, it's not fun anymore. It becomes a fixture, it becomes a business, or it's a trailer queen. A hot rod is something you enjoy. You should be able to drive it, and other people should enjoy looking at it. A hot rod doesn't have to have a blown Hemi and this, this, and this. What's a hot rod mean to me? A hot rod is a car I keep.

12

John Davis (Alamosa, CO)

My hot rodder friend Greg Hrehovcsik introduced me to John Davis in Alamosa, Colorado, in August 2013. In addition to doing machine shop work for all elements of the agriculture industry, John was the go-to machinist for a generation of southern Colorado hot rodders. He could fabricate just about anything. Although he retired in the late 1980s, to this day he is known as one of the area's original rodders and for his sweet-running engine builds.

JOHN DAVIS: My father died in July of '54. He worked at the railroad. Hell, he just had a heart attack. Of course, he was a heavy smoker. Looking back, my mom was only 43; she was a young woman. And she had to deal with us two boys. My brother was 10. I was just 16, maybe 17. And that's when I became the head of the family. In September, she told me that I was the oldest and would have to get a job. My older sister and my brother had already gone; they'd left the house and got married. So I went down to Van's Machine Shop and got a job sweeping floors. And I worked my way up and I finally owned the shop at 38 years old.

I went down and talked to Neal Van Sickle, the original owner. And he said, "John, I'll pay you 50 cents an hour if you'll get here at 3:30." I told him that I can't do that because I get out of school at 3:30. Well, he said, "If you can't get here then, get here as soon as you can and I'll still pay you for two hours." They quit at 5:30. So I got out of school at 3:30, put my books in the locker, and ran to Van's Machine Shop every day. Luckily it was only three blocks. I worked for two hours and made a dollar a day during the week and then on Saturday I worked 8:00 to 6:00. That was nine hours; I made $4.50 that day. My paycheck on Saturday was $9.50. And that doesn't sound like much, but I had a good time. Gas was 14 cents a gallon.

My first car was a '37 Ford Coupe. It wasn't too bad, but it burned a lot of oil. Somebody had taken the windows out; it had a windshield and two little rear ones, but they took the side door windows out and I know why: because it burned oil. If you drove very far in that, you had a fog inside. That car was an oil burner. I was the head of the family, so I took us to school, my brother and I. He says, "John I hate this car. I get out of it and I stink like old burned oil." And he was just starting to notice girls, you know? So I got rid of that one. I think I sold it for $25.

Then I got me a '54 Ford. By then I was a junior in high school. And a '54 was a pretty new car, for a young guy. I've always been a Ford man; I've tried Chevys, but I gravitate back to Fords. Now the knock on Fords is they kept changing things; not like GM where you can mix years and years of the same stuff. Fords are a little hard to keep up with. So I got that one and I used it till '60, I think. No, that was till '57. And that was the Chevy year. A '57 Chevy; I found me one of those.

My first hot rod, like I say, was in '54. I bought that hot rod in San Luis. There was a little body shop down there and the guy who owned the shop said, "I have this hot rod and my girlfriend locked up the engine. I'll sell it to you for $25." It was a 1930 Model A Roadster. With no top, and no seats or nothing. So we went down to San Luis, my brother and I. Did I mention he was 10 years old at the time? And I was making a man out of him. So we picked it up for $25 and towed it cross country on the back roads. And it did come loose in Santa Casho, Colorado. It took off across the prairie and jammed into a sand hill and just stayed there. And then we had our lunch. That was like the old days; we had our lunch before we went and looked at it. It wasn't going anywhere else. We hooked it back up and came on home. That was the beginning of the '30 hot rod.

I was thinking about putting a fiberglass body on that engine. I was going to build a tubular chassis. Everything was brand new in the '50s; tubular work and a lot of the hot rod type things you could buy was either Warshawsky or JC Whitney. Manufacturers were innovative, you know? I think Weiand manifolds were from just a family guy that got started. Anyway, I picked up skills pretty quick from working in the shop. I remember on Christmas Day in 1954, Neal Van Sickle, the owner of the shop, came by and brought me a set of micrometers. They are still over at the shop. I can show you. Anyway, he said, "John, you are going to be a machinist. I can tell. You just have the aptitude." And I said, "No, I want to go to medical school." I took my pre-med in college and got admitted to Colorado University School of Medicine. I went two years and I yearned for the old cutting of the metal: the hands on. So I did quit and most everyone in my family said it's a big mistake. But I've loved machine work all my life, and body work, anything of metal. I just have to build with my hands.

Even when I was in medical school, I would look out the window and I would daydream a lot. I didn't flunk out, but my heart wasn't in it. I went down to the hobby shop and I bought me an engine, a McCoy 29. And then I designed a whole fuselage and everything and built it out of balsa wood. It's just in you; you have to work with your hands. I did this even while I was in medical school; I would work on models in the laboratory where I had a desk, you know? And I'd be putting together a model airplane. It was something to do with your hands.

When I was in medical school, these low-income families lived down where these wrecking yards and the junk yards were. And the children developed this bone malady. And they were burning the old hard rubber battery cases in their fires. And the lead would get in the air and get in the children. And they took us in—I was in the learning phase of my medical education—and they took us in and showed us X-rays. The kids' bones stood out just like a black pencil. Their bones were just vivid because they were impregnated with lead. And of course, the kids were exhibiting other symptoms of lead poisoning. That was a learning process; and I thought, "poor kids." I wanted to be a pediatrician because I did spend a lot of time over at Children's Hospital in Denver. I had a heart for those young kids that did not have a chance. They were dying, you know? And then anorexia was just beginning to be recognized and they took us down—these are the two things that stood out in my career in medicine—and this girl was on the stage and she was only 10 and just a rack of bones. It was very sad. And you still see it nowadays. But it's understood a little better.

Of course, I worked on the hot rods. I built the 324 Olds engine with three carburetors and that's when it was just an engine and I was trying to get something to cover it. And that's when I thought I would get a fiberglass speedster body. But I got dissuaded from that, because I learned that fiberglass would crack early and it wasn't strong. So, I went with the Model A Roadster that I got in San Luis. And that was when I really got started. I built it basically at Van's Machine Shop because he had a back room that he said I could store it in. I'd work on it in the evenings or weekends. And it took me three years to build it.

John Davis displaying set of micrometers given to him in 1954 by Neal Van Sickle, owner of the machine shop that John eventually took over (August 2012, Alamosa, Colorado).

And I built the frame out of a '49 Ford; I wanted the frame to come up and go over the axle a little bit lower. The front axle was a '48 Ford. The engine, as I said, was a 324. And the tranny was a '49 Ford three-speed, which was a weak one, but I just nickeled and dimed it together. To get it all to mate up, I telescoped the Model A frame inside the '49 Ford frame and put gussets all in between. And it still has got the original frame; you can still see that on the car today.

Eventually, my brother went to work for Ford Motor Company and they were teaching him paint and body. He was about 16 at the time, and he painted the Roadster for me in the backroom of our old house. He did a very nice job; he was very adept. And Clyde Pace, a local body man taught me leading. I still stay with leading, although there is a place for body filler and such. And it's very good stuff; the coefficient of expansion is almost identical to metal; which is what you need. But lead seems to be the old standby.

But then again, OSHA is restricting that. But you can still buy the old lead bars, which you can use but you cannot buy lead-based paint. It is outlawed totally.

But as far as getting back to hot rods, my headlight bars came from California. There are four. Remember in the old days they had little fog lights mounted on the bumper? So I got four of those fog lights, but they were yellow. In '58, GM came out with the dual headlights and the bulbs fit right into the fog lamps. I did use the '49 rear end under it. The front end was the '48, but they still had the 5½" hole pattern, so I did not have to carry two spares.

What was I thinking when I built my first hot rod? Oh, I was basically just winging it. I'd see something in a magazine. There was only about five or six magazines, maybe less, that were published in the old days. *Hot Rod* was one of the oldest. It was my favorite. I got ideas from the little magazines. And there were mail order companies that had some of the parts I needed, like Warshawsky and JC Whitney, for the common stuff. Back then I don't think anything was imported. It was all made here. That was the nice thing. We really didn't know much about imports.

Popular Mechanics was a magazine that didn't have much about cars in it but a lot about electricity and small things. They did have a few models in there, because I patterned a boat model and made it from scratch. They told me when I started that it was in three issues—this was the early '50s—that it will float a brick. And a brick weighs six or eight pounds maybe. So I got the hull done and I ran some water in the bathtub and I tested it out. And it floated the brick. It was a bit over a foot long and it was radio controlled and all of that stuff, but I was so poor that I couldn't afford the radio controls. I built all the balsa and wired it, but I couldn't afford to buy any switches. You remember the old dome lights on the car that had a switch on a toggle? It was just a slide switch. I used two of those. But that was as far as I got; wired for radio control but no control unit.

But I am still working on the old metal, still machining. Like I said, I did buy Van's Machine Shop at 38 years old. And I kept it until about '87 or '88. When I took over the shop, I had all the tools to build project cars but I did not have any time. And I had kids, twin girls and two boys. So I didn't get much done other than make stuff for people. But it was a good time. I still get nice feedback about my work. One guy said in the paper that, "I sure hate to see John sell his machine shop because he was the only one that could build anything that worked." And that was a nice compliment. And that was in the newspaper, *The Valley Courier*. At the Alamosa Early Iron car show held every Labor Day, there will be local cars that have parts that I've machined. Mostly the engines; I concentrated on engines.

In the later '50s, three of us started the Rickshaw Car Club. At the height of it there was probably about nine of us. In the club there was a '39 Ford all chopped down, an old Dodge with a big six-cylinder Chrysler engine, and a '46 Plymouth that had flames on it; flames were common then. There were several others. Johnny Rodriguez wasn't a member of our club—he was a member of another club—the Road Gents. He was an original member of the Early Irons Club. There were about three clubs; I can't think of the other one. There was a little bit of a rivalry between the clubs, but they weren't gangs; we got along with each other pretty good.

We also got along with the local law enforcement; we really did. One time when we just barely got my hot rod running, no seats, a piece of brazing rod that went from the dash up to the three carbs I had them operable. Steering but no headlights; I had not got the lights yet from California. And I took off one night; it was kind of dusk, but I could still see. And we were going down toward Lahara on the highway and a patrolman passed me coming down Alamosa We had just got it out of Van's Machine Shop. We'd left the doors open with the hopes of going around the block and coming back. But we had to go out on the road. We were sitting on milk cartons, the old metal wire ones, no backrest. Every time we pulled the rod, it would throw us back. And as soon as we passed the patrolman, he made a U-turn and started going back the other way. And we made a U-turn about the same time and he got kind of stopped with a car coming and we hightailed back to Van's. Which was up to Sixth Street, down two blocks and in the garage. And the guy that was with me, he jumped out as we rolled in the garage and I said, "Shut the doors quick." The lights were off. That patrolman came by within three seconds; and he had his spotlight shining under loading docks and everything. He knew we were there, but never found us. That was one time I escaped. I probably would have had lots of fines on that one. There was no electricity, I had a generator on the engine to get me a water pump, but I had it jumped so it didn't go to a voltage regulator.

Another story on the hot rod. We got it out one night and we still didn't have any lights. We went down to Twelfth Street on Alamosa toward the old River Road. Nile Langston and Garland Parker were the two policemen. They caught us. And they come up, one on one side one on the other. And they said, "Why don't you boys go home? It's late and this car is very illegal. Go on home, it is 9 o'clock." That's really what they told us. And they said, "If you don't go home, then we'll get tough with you." We knew it was time to go home; so we did. Nile died some years ago. I see Garland at the laundry once in a while; he's 83. We have fun reminiscing. They didn't arrest us; they were very nice about it. But they did say, "If you don't go home...." Well, we knew it was time.

Am I still looking for old metal? Oh, I like to; but it's mostly the hunt. I just don't have time to finish them out. I might turn somebody else on to it. I am 75; in reasonably good health. I was an advisor for the vocational school's industrial arts department. I didn't lack motivation or do drugs or anything. I still stayed clear of getting stupid. I had a pretty level head on me for just a riotous young man. But it's a wonder that I survived. I did drive the hot rod from Monta Vista to Alamosa in eight minutes. That's 17 miles. No speedometer on it; that was stupid. But I survived.

I did dig car parts out of the riverbed. Because the wrecking yards didn't sell much stuff. Maybe the frames. But the fenders and even the whole cars were dumped in the river channel for riprap. They lined the cars upright and some were just thrown in, so it's wherever they landed. But there is one arroyo where they lined up these cars from the '50s. And there is a Crown Victoria upright there with so much chrome on it that if the sun is just right, it looks like a gem sitting in the side of that arroyo. We used to go down there to get parts. But it's probably gone now.

My wife owned a Crown Victoria with a continental kit. And we sold it for $200 when I got admitted to medical school. Her car was a factory continental kit. Those are rare. The bumper matched the kit. Are there any other cars that I regret selling? Yep. The '57 Chevy. That was a good car. I say I am a Ford man, but that '57 was a good car. It was a two door, hardtop; the most likable one. You know they are using posts now and even four-doors. You'll see some at the show. There are companies that are making the Tri-Five car bodies. I think they are doing that because people of all generations have fallen in love with those cars. The mid-'50s really seem to stand out. You've got the '32 Fords and up to the war. And then it kinda drops off. And then the '47s and '48s; they are kinda getting popular. And then the '50s. It makes me feel good to see that. That I can communicate with some of these young kids, as far as the '50s cars. I can't communicate with them on the early stuff.

The 350 Chevy is the choice, because of the availability of the parts and the interchangeability. You can build a pretty nice 350 for $3,000. And that's pretty good horsepower too. And I tell everybody if you want a cheap hot rod you get a 350. That's my motto. And with the Olds valve covers for the 350, you can make it look like the Rocket. My old Rocket, when I had it dressed up.... Well, I wish I had picture of the engine. It was beautiful. Because it said "Oldsmobile Rocket" right on top of it. And the spark plug wires came down and they went right to the spark plug and it was just uniform. I put 12 volts on it and because of my compression, I planed the heads, I think I went .125 on them. But then you interfered with the manifold, so you had to plane the sides of the heads to bring it all down.

But Old Van Sickle, he was an old man who I don't think finished high school, but he was totally metal. He would get a picture in the mail of a new machine, a surface grinder or something like that, or a rod reconditioner. And he says, "We can make that." I helped him make a few machines. He had a surface grinder that he made from scratch. And it's still in Ben's Machine Shop. And it works great. And when he left Van's Machine Shop, he had two tables cast at the Pueblo Foundry and they were about 500 pounds apiece, because they had ribs to remain straight, and he took one when he went to California. And he says, "I am going to build me a boat and I want you to build me an engine." He was a nice man. He says, "It will be my last engine. I want one of your engines in my boat." That was a '55 Chevy 265. And so I built his last engine.

There's so many engines I built. I built Eric Christenson's first engine when he was 16. And it disappeared into the Valley somewhere. He tracked that engine down at 40 years old and bought it back because he liked my engine. Now that's a compliment. But Van Sickle took the '55 engine and built him a boat; it was a 24 footer, or 28 something, it was a pretty big one. Fiberglass, all hand laid. And he took it down to the shore. And he finally told me, "I just can't find a place to park the boat." He finally sold it; I think he got it in the water once. We even built the water-cooled manifolds at Van's shop, because he didn't like to spend money. We built them out of small tubing for the exhaust and then bigger tubing for the water jackets. It turned out nice; I'd like to have pictures of those.

Who does this kind of work anymore? Well, Van's is doing quite a bit. They've got CNC machines that do far more than I could. Now, if you can imagine it, you can make it. They are just down the street from here. In fact, Steve at Van's built the 351 Cleveland for Greg Hrehovcsik's Mustang and a 352 for his '66 Ford pickup. Both of those engine just run like tops. Are the kids learning these skills? Well, they really are not. I tried to teach the kids at Southside just simple lathe operations or threading or types of threads and stresses and stuff. And they say, "It really looks nice." But I don't think they could make one.

I tell you what; at the shop there was Van Sickle, myself and one other guy—there

John Davis and Greg Hrehovcsik standing beside John's '49 Mercury Woodie Wagon project car in August 2012 (Alamosa, Colorado). He sold the car to Greg in 2016.

was only three employees. And one day Van Sickle and the other guy was sick; I was the only one there sweeping the floor. I was doing only menial jobs. And his wife was there in the office doing book work. And a guy came in and wanted a thread put on a shaft, which is a simple operation, but I had never done it. Mrs. Van Sickle came out and said, "John, this man needs this threaded." And I said, "I'll try it." I'd watch them do it. That's all I'd done. I'd watch them. And I knew how to chuck it up. And I had his threads cut, and I didn't screw up. And the guy was very happy. And that was one job that I remember, because I winged it totally. I knew how to engage the thread dial—it was all hands on in those days—if you turned it to .002, it cut .002.

Nowadays it's all programmed. So that's the nice thing—you see an old lathe with all the chrome handles—it's nice to know that you could run one of those. But those are disappearing. Well, with a CNC, if you want 10 or more items, you can do it. But you can't make just one because your programming takes up 99 percent of your time. Actually, when you punch the buttons, the other one percent makes the part. All the time is in the preparation. So you need to make like 100 parts.

The first CNC that I remember, as far as simple part reproduction, was a hydraulic tracer lathe. You'd make a part by hand and then you'd put it in the machine, set the dials, and with the part in place, these hydraulic fingers would follow the part and would cut it out. That machine is still out at the shop. It was fun to watch them because those fingers would feel it just like a person, and transmit it to a sensor switch, which is simple. They don't use them anymore, it's all electronic resistance, I think. But anyway, that was nice.

And that would reproduce your part. But there was still a lot of room for error. If there was an air lock in one of the lines, it would screw it up. But it was a step.

So much has changed. I told you the police story. Back in the day, the police—local and state—they treated us nice. They treated us like humans. A lot, any more, they treat you like a criminal before you go to court. They are revenue generating; it's greed. I know there's a reason behind it, but some of them just make me mad from the word go. And you just can't resist them, because you're in the system from when you are born. You just can't resist it. You can get mad and talk about it, like you and I are talking, but there's no compassion in a lot of society. Even the medical field, it's kinda lacking compassion. If you go into a doctor, they steer you to the financial adviser. It's always upfront. They still ask you how you feel and do you want to sit down. But then here's the clipboard; start filling the payment forms out.

I know there's more stories. I got a lot of project cars still. I got a '49 Merc out at Christenson's yard. And I have another one on the other side of town, at the old honey house, where my dad fell through the roof. A month later he died of a heart attack. I think something shook loose; he was 47. I guess I got about eight cars out there.

13

Bob Austin (Rochester, NY)

Talk with Bob for a few minutes and you know that he is an original. He's been a hot rodder for more than 60 years. He and his brother had their Model T roadster featured in a 1959 issue of *Hot Rod* magazine. He still builds them. And he does bodywork and paint. Bob drives his four-door Phaeton all over Central New York. He told me that he and his wife just jump in the car and go: "We never know where we are going, we just go."

BOB AUSTIN: I think I inherited the mechanic thing from my Uncle Charlie. He just died about four or five months ago, at 94. My uncle was quite a mechanic, the guy could fix anything or work on anything; he just understood it. And my dad was pretty good too. I was a mechanic in the Army during Korea. I worked on jeeps and Deuce and a Halfs [6 × 6, 2.5 ton cargo truck]. I just loved mechanic work. And when I came home in '54, my brother Carl had a garage. We started working together in the garage, and people started coming around, shot the breeze, and we let them use the machinery there—the band saw, the drill press, whatever they had to use to keep things going for themselves. And we built a very good rapport with a lot of people. Eventually we moved into a bigger, better garage.

Everybody had a lot of respect for my brother and me. We had a community garage. You don't find a place like that these days. What I liked about it was the camaraderie. Our friends would bring us stuff to use from where they worked. They used our stuff and one hand kinda washed the other. Nobody had any money back then. It was just a matter of being in it together. One guy worked at Xerox; when someone left the company, they would take his whole toolbox—which was given to him by Xerox when he got the job—and they would actually chop the tools. And so our friend, being a semi-boss, would get a hold of some of the tools before they got chopped up. And bring us some.

My brother bought this Model T from a local junkyard. It was just the roadster body. And it was a pickup. That was the thing; it was a model 27 roadster pickup. And I got out and I said, "You know the pickup's nice, but a roadster with a turtle deck would be better, if you could find one." We did find one in Olin, Iowa. A buddy drove us there because neither of us had a license. A week after we got the turtle deck back, the guy that drove us got killed in a car accident. He was just going to Batavia for a cup of coffee after work one morning.

FoMoCo FLASHBACK

With a few novel exceptions, the Austin Brothers' roadster is a nostalgic blend of the Ford Motor Company's best

Rochester, New York

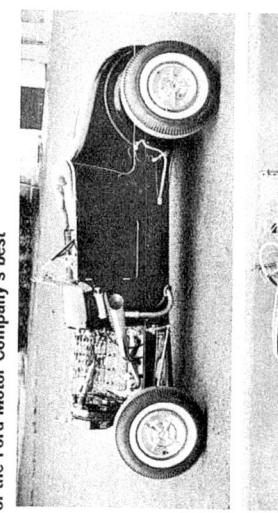

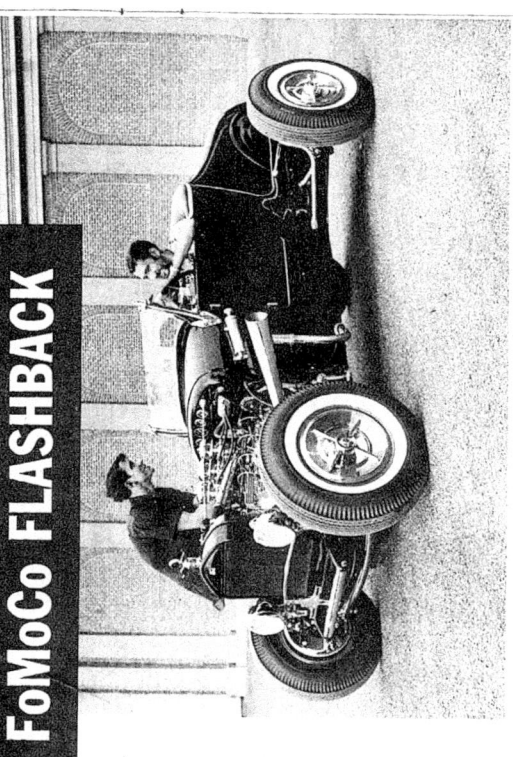
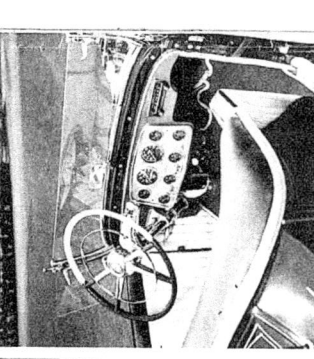

LEFT — Austin Boys took 4½ years to assemble this show rod, using a '27 'T' roadster body on a '31 'A' chassis. Loaded with chrome goodies galore, the car won 12 trophies out of its first 7 show appearances. Competition minded boys also run drag bikes.

RIGHT, ABOVE — Low windshield, with windwings, consists of lower half stock 'T'. Headers are inverted 'W' type, megaphone baffle visible in photo below. Frame from firewall back is encased in bellypan. Rectangular roadster turtle deck. Note wheels are from Imperial.

Slight rake is attained with 9.50 rear tires. 'T' was purchased as pick-up, required trip to Olin, Iowa by Bob and Carl to unearth taillights, contrast nicely with the round exhaust openings in bellypan. Note header baffle.

Bored 1969 over, '48 Mercury retains stock cam for smooth town work, has Edelbrock heads, dual manifold for a little extra jump. Harman-Collins dual coil ignition, and Weber clutch complete line-up. Pressure pump is strictly a reminder of a long-gone era.

FAR LEFT Brakes are '41 Ford. Bell tube axle mounts 14-inch Dodge wheels, 7.50 tires. Monroe shocks control spring action. Royce Motometer, a rare thing, caps genuine 'T' radiator. The steering is '37 Ford, erected on late spindles through Bell arm.

LEFT — Hudson electric shift controls '39 Ford transmission. Steering wheel from '55 Ford, Stewart-Warner marine panel dress interior. Upholstery in red, white leather by Jerry's Top Shop in Niagra Falls. Compact radio is by Cartrol-Custom.

Photos by Xenophon Beake

So my brother and I built a T with the turtle deck on the back. The engine was about a '48 Ford flathead. We extended the bottom of the turtle deck with sheet metal and painted her candy blue. Well, a guy from the local paper came along and wanted to take a picture. He got it into the paper but they called it a Model A. That's the only bad part. But we got a kick out of that. We had a beehive oil filter and gauges are inside and outside. Our car was featured in a 1959 issue of *Hot Rod* magazine.

If you look at the article you see the flathead in it. Later on we put a Chevy in it. The long pan off the back and the side things are from a Model A, inverted to go down acting as a belly pan. It went down from the frame to the running board. We put a fuel pump on the driver's side. My brother loved sprint cars so much, that he just had to have it on there. I did not especially like it, but he had to have it so we went with it.

The tranny on that car was really special. It's a Hudson Select. It's got a short shift stick, but you never touch that. The solenoids did the shifting. Move it into first, second, third, and fourth, or reverse. One solenoid pulls it sideways, so one goes up and down in reverse and first. And the other one goes up and down in second and third. But the other one makes it go crossways; it's called pre-select. You put that little lever where you want it and push your clutch in. Automatically it goes right in there. And then while you are screaming out in low gear, you put the little lever into second, but it won't go into second until you hit your clutch. As fast as you can hit your clutch it changes gear. The shift lever jumps forward. And the same thing for third. I still got some of these '35 Hudson electric shifts upstairs in the garage.

My uncle figured this all out. We knew he had it in a '36 Ford. And so we went up and we hunted up that '36 Ford. And we found it down near Cortland, in a junkyard. We called the guy, and he says, "Yeah, I had that for quite a few years. But we finally junked it; we wore the hell out of that thing." And so we went and saw my uncle, and I said, "We want to put one in our roadster." And he kinda helped us out. And we started going to junkyards finding, I think it was, '35 to '39 Hudson parts.

Now back in the day, you could go to a junkyard and you could get decent stuff, good old metal. You can't find that today. In fact, a lot of them won't even let you in. We had one here, on Town Line Road, it was called Bucky's. Buck would let you go in. But then you know what happened? Some people started picking up parts and throwing them through fricking windows. And that kinda killed it.

Years ago, if you were driving a hot rod that didn't have fenders through Buffalo, you were going to get a ticket. Our roadster didn't have fenders, so we always told ourselves, "Never go near Buffalo." But we finally did; we entered and won the Buffalo Auto Show. One time we were at the Hartford Show and we met a guy with a Model A roadster that was channeled so far over the frame that if the thing flipped over the wheels would still be on the ground. We lost the show to him. The judges wanted us to open up our trunk. And all we had in there was a vacuum tank for the Hudson electric vacuum shift. It was all neat and everything. But we didn't have a fire extinguisher mounted, we didn't have a first aid kit mounted and so we did not win "Best in Show." That kinda got me mad because I didn't come to show off a first aid kit or an extinguisher. I came there to

Opposite: **February 1959** *Hot Rod Magazine* **article on the Austin brothers' 1927 T roadster body on a '31 A chassis.**

show my car. I don't want to open the hood or the truck of any of my cars; you are looking at the car, its profile, door gaps, and the finish.

The last time I drove it, I was going down East Avenue here, which was the ritzy section here, and some guy was just poking along. This really stirred me up at the time, he was poking along and I am coming up to an intersection, and the light is green and he's just poking, so I pulled out, tromped on it, that thing just hooked up. I went around that guy so fast, tires just spinning, and I was sideways, and I straightened it out, and thought, "Holy Christ, why did I ever do that?" Well, I had the green light, but it was through the intersection and that kind of scared me.

My brother was a year older than me; I was 67 when he died. We got along basically good. A year after he died, I took two wrenches on a hunk of cable, and pinched the ends so nobody could steal them and I draped them over his tombstone. And when I was walking in, the guy that owns the graveyard saw the wrenches. I had to stop and tell him that my brother and I wrenched for years and I am just hanging a couple wrenches up in case he wants to use them.

My brother's youngest boy has got the T roadster now. He owns it, but it's all apart. How he ended up with it is a story. My brother and I built it together; it was a partnership. And he couldn't drive it very long because his legs were too long and he'd get a cramp. So he'd have me drive him. Anyway, my brother came out here one day and had a brand-new motorcycle and he said, "I want to trade you my brand new motorcycle and $2,000 for the roadster." And I said, "I don't want your 2,000 bucks and I don't even want the motorcycle." But I said, "I'll take the motorcycle if it will make you feel happy." Well, he said, "I gotta give you something for my half of that." Well, he didn't have half; he had seven kids and he couldn't even afford to get it built. But I wasn't going to be a prick about it. So I let him have it. He took it all apart, and was starting to redo it. But he didn't like bodywork; my brother hated a hunk of sandpaper in his hands.

Well, after my brother died I didn't get the car back; it was in his basement, in his wife's care. All the hot rodders that hung around the garage said, "Go tell Kay that you want the car back because you want to put it back together while somebody's alive that knows how it goes back together." And I said, "No, I made a deal with my brother. And it's hers to do what she wants with it." They say, "Geez, it may never see the road again." I say, "I don't care, you know?"

Like I said, my brother hated sandpaper. He had these two '34 sedan doors that he was working on. The doors needed work because we had beat the shit out of this thing for years. Anyway, he's got these two doors on the sawhorses. And he says, "Hey, hi Austin." I say, "Hi boss, what's happening? Got your doors all ready?" And he says, "I think so." So I say, "Can I feel them?" You don't touch nothing. He didn't like people touching them with greasy hands or nothing. He says, "Yeah." So I wipe my hands down and I started feeling.

And all of a sudden I see him staring hard at me. It's not right and he knows it from the look on my face. And he says, "What's the matter?" I say, "They are not right yet, boss." He says, "Aw … come on." I told him they are not ready yet. So he comes over and he starts feeling them and he says, "I don't feel nothing." So I look at him and say, "You know why? You are using the wrong hand." So he looks at me like I'm an idiot. But he starts feeling with his left hand. And he's feeling and I can see the look on his face change. He says, "Jesus Christ. They are not right." So he looks at both his hands and he says, "Why is that?" And I told him, "I don't know. But I can tell with my left hand if a bump is male or female, but with my right hand I feel nothing." Try it sometime.

So I asked my brother, "Do you want me to work on the doors to get them straight for you?" I told him, "Where you made your mistake is when you got so many little ripples you don't concentrate on them. Put a new layer on the whole damn door. Then bring it down." I told him to get the primer off and get it back down to fiberglass. These were pretty good-sized doors; you cannot concentrate on the dents one at a time. You have to get the primer off, put another a layer on the whole door, start blocking it down, and curve with the block. I told him everything will come right. And it did.

But my brother Carl knew lacquer. He was real good shooting lacquer. On my Phaeton, I did the body work and he shot it. It came out like you were looking into a mirror. We put on four coats, we would wet sand it with 400 or 500 grit paper, and then he'd put on another three coats. And when he put the second set of three coats on it was almost like glass. He would spray that lacquer so beautiful. But for his car I said, "You don't want lacquer. We are going to go with the black urethane. I'll show you how to paint it; then I'll show you how to rub it." You know, if you do it right, it's almost a shame to rub it sometimes. That's how nice it comes out.

Well anyway, he painted it with urethane. But he's real meticulous. Around the areas where he knew he might rub through he used the old-fashioned buffing pad, not the new foam block. I don't like the foam block either; I am the old-fashioned guy. I like the wool. So I go number fives, and I've never had bad results with that. He would very carefully hand rub around every curve, because a '34 has got some real delicate little lines on it. And he did not want to rub through.

I had to show him how you can rub out urethane. And I got on one of the big quarter panels. I wet sanded it with 2000 with soap and water. You have to use soap and water because the grit will load up in your 2000 and you will scratch it. So I did that, and then I said, "Now I am going to show you how to rub." And I rubbed, and I was pushing so hard that I was almost bending in the metal. But on a '34 that's pretty hard to do. And I kept doing that. And I said, "See that? You cannot hurt this paint." And I kept doing that. And he said, "All right, all right, I got it." I did it just to piss him off. And then he finished rubbing his car out. I'd straightened him out.

Man, am I a nut about finishing a car off right. I won't let anyone help me sand or nothing. I am doing the work and I am no slouch. I don't know, maybe I got a knack that I can feel with the sandpaper how it's going whether I got it down as smooth as I want it. I am not afraid to overdo it if I have any doubts. I sand from the top down so the dust is going to fall down where I've got to sand anyway. And when I paint a car, I start at the bottom of the door and work up. I don't start the top. I never let anyone touch a car I am sanding on. Everyone has grease on their hands.

I've worked on a lot of cars for people. But if you build one for yourself, you don't necessarily want to sell them. You don't. I am funny that way. I tell everybody, "I got the same wife, same kids, and same cars as I had when I got married." Like my '34 Phaeton; we go everywhere in it. I rebuilt it and we put a Chevy in it, different wheels, box frame, big Ford rear end, 9" in the back. About a 3.78 rear gear, the motor really winds pretty good, I like it. Engine is a 350 Chevy, with a cam in it, and a 350 turbo tranny. I painted it about 45 years ago with black leaded lacquer. But it's starting to crack on me now.

Bob Austin with his '34 Ford Phaeton. He painted it with black lacquer 45 years ago; it needs to be repainted but he's going to leave it alone because he's "got enough other work to do without worrying about repainting" (September 2009, Rochester, New York).

Fatigue crack like. It needs to be repainted again, but I am going to leave it alone. I've got enough other work to do without worrying about repainting that thing.

Somebody is making a Phaeton body now. So sometimes people ask, "Is that a kit car?" I tell them that my car is a four door. That makes it original Henry Ford. And I tell them, "Henry built it and I keep fucking around with it." Excuse my language. My youngest son gets this one. I've built a '34 pickup that I gave to my oldest son. I chopped it five inches; otherwise it'd look like a telephone booth coming down the road. Cut it five inches, straight down; you can do it. And I built a '32 three-window coupe that I gave to my grandson. I finished the '32 about four and a half years ago. My wife and I bought it a week before we got married, and that was 50 years ago. And I just never touched it, all them years.

I told you my brother liked sprint cars? Well, we raced supermodified at Oswego for nine years. Lee Osborne raced the car. I raced it once; I thought I was flying until the slowest guy up there went by me like I was standing still. Anyway, Oz builds a good car. I painted a few Brookville bodies for him, you know? I know of two that are around here. Anyway, I've painted a number of them. You know, my brother always said that there are more '32 and '34 Fords on the road today than Henry built. Thanks to Brookville and guys like that.

Brookville's stuff is pretty good, but not perfect. On their roadster body, where they come with that rib down over the trunk lid is just a bit off. Same on their three-window coupe; it's not a nice smooth fit. I told them I found 14 places on every Brookville body I've done that needed glasswork. Each side of the cowl vent, where the cowl vent that opens, it puckers there. Now I don't know how they would have done it different, but right there was a major flaw. And down each side in the back, and then right across the trunk lid or rumble seat lid, whichever you had, that shelf there, it was deeper in some spots than others. So I had to make repairs before I painted them. Now Lee told me, he said, "Leave the fricking things alone; just paint them." Well, I can't do that, I can't do that. You know, it's got to be right.

I did one for a guy who is down in York, Pennsylvania. It used to be an old drag racing roadster, and the frame was so weak that I could press on it and I could watch the roadster doors raise and lower. That's how weak the frame was. So I said, "Why don't you do it right—box it and everything?" He said, "No, no, nostalgia. You ever hear of nostalgia?" I said, "OK, nostalgia." Back in the day, you could throw a car together or you could build a car.

Probably the nicest guy I worked for, doing a car, is a black guy by the name of Ken Wiggins. It was a '50 Chevy coupe two door with lots of stainless moldings. He wanted me to paint it for him. After I stood back and looked at it I told him, "Ken, I don't like all this stainless on it, makes it look like a patch quilt." And he says, "I am open to whatever you want to do." He said, "I have no idea of what I want, but if you want to do something that you think will bring attention, do it."

So I left all the stainless off, had to weld up a bunch of holes, where the clips would go in. I got it pretty much ready for paint and then I stood back and looked at it, and looked at it, and I thought, "Man, that's ugly. That's an ugly car." The top looked like it had a forehead on it. It was really, really big from the top of the windshield, up and over. Man, it was just thick. I called up Ken and I said, "Ken, I've gotta chop the top." He said, "Well, I don't want you to chop my windows." And I said, "I am not going to touch your doors, I just want to get rid of that high forehead look. It's ugly." And if you've ever seen a '49 or early '50s Chevy, that's the way they are.

Ken said, "Do it." So I chopped the top off, three inches up, and then cut Vs in the metal and bent it all down in, to line things up. And then I laid the cut-off top on the floor and started walking on it to stretch it. Thank God I am not a heavyweight. So I walked on it and walked on it to stretch it and to bend everything down. Then I set the top back on and I welded it all the way around. And now, instead of being ugly, it was lower. And, for some reason, a lot of people catch it, and they like it.

Anyway, I have worked for a lot people, never had any bad reports, nobody give me a hard time. But Ken was the only one that told me to do his car as if it was mine. He actually said, "I want you to build the car like it was your own." And that's what I want to do and that's what I did. He was so willing to go along with everything. So the top on his car was the original, but with me walking on it to just stretch it enough. And I didn't dent it. That was the amazing part. I thought sure as heck, I am going to have some ripples or something. And I glassed around the whole thing, painted the car black, and it come out gorgeous. That was a good project.

I have a good time when I go to shows. But you know who raises hell at shows up here? The Canadians. They come down to our shows and, Holy Christ, they are nuts. They are nuts. It's like they are in another country and they figure they can't get in trouble.

And they are going to go back home before they pay a fine anyways. It's kind of a "so screw you" thing. I really like some of those guys.

Anyways, there's a lot of good people in this car thing. But there are some guys who want you to do work for them that have a bad reputation. They want the perfect car. They will never be satisfied with anything you do. Local car guys know who these people are and will tell you that you don't want to do work for certain guys. And I don't work for them because you will never do it good enough. You are going to laugh, but I got a big mouth; I know that. I am not afraid to express my opinion. I've had a few people say, "You still got all your own teeth?" Because I say what I am thinking and I don't give a shit.

There's a high-end thing that I don't like about the car deal. A lot of my buddies say, "How come you didn't come to the show the other day?" or "how come you don't come down to Rick's which is tonight?" Rick's Place has great prime rib. They get about 400 cars; it's a big thing every Monday night. I've only ever been once there. You walk in and, everybody asks, "How come you didn't come to the show?" I just, well … you never got anything as good as they've got. You haven't got as much money in your car, and so on. You know, some of them just piss me off. We are there to have a good time to talk about cars, how things went together or whatever, and it's not anything like that. A lot of these guys had someone else build and paint their car.

I got my Phaeton by trading a '34 Ford truck for it. The Phaeton body is very special. For some reason, the older the car gets, or the older I get, I stand back and look at it when I am coming out of a restaurant, and it's so pretty. Just enough rake, but not standing on its nose. I hate people who set their vehicle standing on its nose, just to make it lower. All the weight is up front anyway, so why put more on the axle?

I like the '56 Ford truck, except it looks like a walrus. They are just too big and heavy in the front because of the big bull nose there. Who is the guy? Chip Foose, right? He did a '56 pickup; really nice job on it; he brought the hood down, made it look good.

Bob Austin's fleet of Fords at a Rochester, New York, area car show. From left to right, a '34 pickup, a '32 three-window coupe, and a '34 Phaeton (June 2008, photograph provided by Bob Austin).

I often wanted to write him and ask him if he could give me directions on how to do that. Send me some pictures. Not that I'd want to seem that I wanted to get something out of him. I just would like to have my truck looking like the one done by Foose. He is supposed to be a really good guy. I heard a story about him coming up to a guy at a show who was trying to adjust his pickup doors. Foose grabbed a wrench and they got the doors hanging perfectly after an hour.

I did a 3-inch chop on my truck cab. You know, they tell you never over 3 inches, because of the wraparound windshield. That's what happened on my '56; I busted five windshields. Well, I busted three and two different glass shops busted the other two. One guy in the glass company brought me over a new window grinding tool and says, "Here you try it." So I started up, it fed oily or soapy water to cool the glass down as I went across the top with the grinder. Naturally it was a battery-powered thing, and it ran out of juice, so I let it stop and set to charging up. Well, I went back and looked at it, and where I stopped, it cracked. And nobody moved it or touched it or nothing.

Then I tried one using duct tape. I left a little 8-inch groove like that all the way across and down, taped the snot out of it, and sandblasted. And that works. But you get impatient. You start getting closer to finishing it and then heat gets it and cracks it. So I tried three of them, and finally I said, "No more." They were a couple hundred bucks apiece. So five of them gone, that was over a thousand bucks. No luck. Somebody did tell me there was a guy, supposedly in the Webster area here, who was starting to do them with laser. Maybe that works; I don't know. I finally settled for a piece of Lexan. Heated it and curved it, got it to curve right. It gets marked up and I just make a new one, that's all.

When I first started planning out my '56 chop I found an article that said to cut the top off and then cut it into four pieces. Then you can just lay another top over it and weld all the way around. I thought, "Man, that's a lot of work." So I went out looking for a donor roof. I saw a '56 cab in a guy's backyard. So I pulled in and asked him, "What are you doing with the '56 cab out there? He said, "Nothing." I told him I wanted the roof. He said, "What are you doing?" I told him that I was chopping the top on my '56. I explained what I needed to do. He said, "OK. I could see that working." Then he says, "What do you got in it for a motor?" Then I told him a Chevy. Holy mackerel, you never heard a guy blow his stack. He told me, "Get the fuck out of my yard, don't ever come back. I cannot stand people that don't put a Ford in a Ford." He went bonkers on me. So I had to go away without it.

Eventually I went without the donor roof. I stepped back and looked at the cab and I saw that the way the sides come up and that the back was pretty darn straight. I thought to myself that I shouldn't have to cut it into four pieces. Two should be sufficient. So I cut the roof dead middle right straight back, chopped three inches out, and brought each side down. Everything welded up perfect. Perfect. But I had about a 2-inch gap in the middle of the roof. So I cut a 6-inch strip of metal, I bent it around, and welded it to the inside of the roof. I glassed that up and when we got all done, the thing turned out beautiful, and everybody was happy. And it's still that way today, and it's got the plastic windshield.

So I did it my way. My '56 is at my son's house right now. But nobody drives it. My son hasn't driven it for about three years. And what's bad about this state; if you haven't driven your car or registered it in three years, it's another 50 bucks to get a new registration, you know? They got you coming and going.

14

Rick Treworgy
(Punta Gorda, FL)

Rick Treworgy is the creator of Muscle Car City, a Florida museum that displays more than 200 vintage muscle cars. It is a world-class collection of classic cars and hot rods spanning 35 years of makes and models. The collection focuses on GM performance cars—Camaros, SS Chevelles, El Caminos, Big Block Impalas, Pontiac GTOs, Oldsmobile 442s—and also includes every Corvette from 1954 to 1975. For a Chevy guy, this is a gearhead's Shangri-La.

RICK TREWORGY: This all started by the time I turned 16, when I could legally drive. I'd have a car that was my own and a car that was a project just about all the time. I had a day job and I'd work until like 9 o'clock at night. And then I worked at a gas station for an old man who was a real good friend. I worked on cars as late as I felt like being there. It became quite late some nights.

I worked on a lot of cars that I sold pretty quickly. That was like your gravy money while you were going to school. Your buddies were buying these projects. About the time you'd get them in primer and the mechanics were right on them, they'd want to take them over and do their own paint and stuff. And I just found that it was a niche that I could pick. It started out as an extra income source; it was something I loved doing. And I never really stopped. I kept pyramiding the cars; next thing you know, I'd have three at a time. I've never liked driving them around if there's a spot of primer on them; to this day I won't drive a car if it's not finished. It's got to be done for me to put it out on the road, you know. Some guys that were buying them when they weren't done kind of liked that because they weren't around town yet. And they were the ones bringing it out. If I did it, it wouldn't come out until it was finished. I might drive it around the back road, but it didn't go out to town unless it was painted and looked like a car. I just don't like a car that looks like a project.

My dad had an International Trucks dealership. I went to work in that dealership when I was probably 12 or 13 years old. After school, I'd go there to work. Working for your parents was never easy, you know, but it was an opportunity to be around all the equipment and stuff that I had no way of getting my hands on any other way. And Dad was pretty good if you wanted to stay there late to do something on your own. All he cared was that everything better be straight in the morning. That afforded me the opportunity to always have access to spray equipment and body working tools and stuff like that. I think that helped my interest in the old cars quite a bit.

My dad was also a used car dealer. We'd go to auction with him and kinda bend him into going into the high-performance cars, which at first he didn't want any part of. But then he figured out they were selling and he'd like them a little bit better. That's where I got the focus of the museum. Basically, I collect the cars I liked when I was 16 to 24: muscle cars from the mid-late '60s to early '70s. The cars I liked then are the cars I still like: high horse, 4-speed, and General Motors is what I concentrate on. I want the cars that the average guy could have bought at a Chevrolet dealer and if he was smart enough to check the right option boxes. I am not a big fan of Yenko, Nikki, or Motion Supercars. And I like real matching number cars. I have hot rods, but I always look for muscle cars.

Do I have a car out here that was one of the early ones that I kept? I probably do. But when I started this, cars were leaving me rapidly because it was a money source. And that went on even after I started collecting cars; I still was trading cars around for a few years. I wouldn't let my numbers get below a certain number, but I don't have a core car that I kept from the time I was 18. I just don't. When I started I just was kinda buying and selling cars. I have done that my whole life. I still buy and sell cars. But eventually I started keeping some, and those became a core of probably 20 or 25 cars that I have kinda held around. But even those I've culled over time. If I find one better than what I had, I will buy that and sell mine. I just don't keep up with how long I've had a certain car; I just don't focus on that. If I tell you something was five years ago, it might have been ten.

Why Chevys? My dad was a Ford guy growing up. He became an International Trucks dealer, but he always drove Fords. Everything was Ford, back in early days. I remember maybe '55 or so as I started paying attention to cars. And he'd buy two new ones every year. He'd buy a station wagon for him at work and then a convertible or hardtop for my mother. Ford was the only thing as far as he was concerned. Well, I bought into that until, I think, my second car. I bought a Ford and I redid it and got it all done and I couldn't sell it. I figured out that the parts availability on the Fords was a little bit harder. Ford "had a better idea" more often than Chevrolet, so if you take a distributor out of a V8 Chevrolet and put in almost any other V8 Chevrolet, it will run that car. Ford won't do that.

For parts availability, and ease of getting power out of a small motor, Chevrolet wins. That's proven time and again. You don't see a Chevrolet hot rod with a Ford motor in it. But you sure see it the other way around. It's the parts availability and how easy it is to get power out of them. It was easier for me to sell them when I'd get projects done. But then I'd just fall in love with them. We were racing them and playing with them, and I just got to where I knew what it took to make one of them tick. It's a lifetime love; it never went away.

I have no Fords in the museum. Well, I've got the one truck that pulls Miss Budweiser, and that's it. Everything else is Chevy or powered by it. I've got a couple old Willys out there. One has a big blown Chevrolet motor in it and a Willys jeep that's got a big block Chevy in it. But I am all General Motors. I've got Pontiac GTOs, Olds 442s, and beyond that it's pretty much Chevrolets.

How do I explain the passion? A car guy will get it, and everyone else won't. I don't

Rick Treworgy with the '55 Chevy he is building. As he puts it, "I just like the '55 Chevrolet ... it's a solid enough platform you can overpower it until hell won't have it, and it will take it. It's the perfect platform to build a rod on" (March 2010, Punta Gorda, Florida).

know how to explain that myself. It is a passion; it gets developed in your younger years. I started collecting for the simple fact that I'd always worked for myself. I've been in the construction industry most of my life. And there wasn't such a thing as a 401(k) back when I started. I just went, "Well, if I save a couple of these, it is money in the bank." I knew an old man that sold off some Model A Fords and he was something like 60 or 70 years old. They held their value. I thought, "That's not a real bad idea." Because it's a retirement savings plan.

Part of my passion for cars comes from being a kid in a small community. That helped. I know people that were kids in a big city that didn't have cars until they were 22 or 23 years old. In a small town, by the time you're 15, that's the number one thing on your mind. I can remember thinking about my driver's license at 13 or 14, you know? Your first car was your rite of passage. That's what it meant to us in those days. And it was a fun time for cars, let's face it. I started driving in 1965. There was probably never a more progressive era than the 10 years after 1965. For power it was going off the hook. The things we were putting together we thought were hot rods, GM was building almost as fast as we figured out how to do it. It was a great time for a car guy. And that never leaves you.

It's about nostalgia. It was a golden age. People come into the museum to relive it. Or just for the fond memories. Nothing else brings it out in people. You wander out in my museum for an hour and it takes you over. I mean, it happens to everybody, not just us car guys. There's something about those times that were fun. And they were simpler. Yes, we did drink a little beer back then, but we didn't have the "big trouble" things back in those days. It isn't the same today; it was a lot easier time. I don't know if the world's got any better since then; it's definitely got richer for some, but I don't know if it's got any better.

There are people who look at hot rodding and think it's not good for the environment and say it has no future. Well, I think if you watch hot rodding, your best and most valuable hot rods are built with your best and newest stuff coming out of Detroit. That's what is being put in these old bodies. There's probably more being learned by people on these hot rods than any other group shy of Detroit itself, or Toyota, or whoever it is. There's probably more being learned from these hot rods than from the new manufacturers. And that stuff has got to be a billion-dollar-a-year industry.

I don't see hot rodding going away. I might be wrong, but I think that if they find a way to run clean fuel, they are going to run clean fuel in the hot rods. The bodies are going to go on even if the running gears don't. I mean, take your Model A Fords and the Model T; yes, they are kind of gone and there's a few people that will always keep them for their historic value. But there's a bunch of people that will hot rod them. When you hot rod them, you are putting the latest kind of technology in the car, so I don't think it is a bad thing for the environment.

If you put an LS flex-fuel engine in your truck, who can say that's bad for the environment? And there's hot rodders who are learning to do that. Yes, there are always going to be some interested in flat-out performance, and getting all you can out of it. But they are not a bunch of yokels. Most of your best mechanics in the world turn out to be your hot rodders. Go around these big shops, and you have one old guy and eight young guys working. In the back of these shops are the young guys' project cars. They may look like hell, but by the time they get close, somebody with money buys them and finishes them. And you can say they are going away, but they're not. There are a lot of 20- and 24-year-olds out there that will outwork me all day long. And they are craftsmen. You can think in your mind that you are better than them, but not true. It's an ongoing process. You'll find young guys who are good body men. They are one step from an artist. There are a lot of young men out there doing amazing stuff.

One thing that's different today is that guys aren't using lead in their bodywork. There's nothing wrong with plastic—it's a better medium than lead. The difference is you can't build structure with plastic. You can fill something you've already got solid with plastic, and it will last forever. But you can't hold something together with plastic—that won't work. You have to have your bodywork solid. The nicer your metal work is on the body, the nicer the project will be in the end. But there's a place for plastic and it will last longer in the right situation.

If I were to head out on a Power Tour, what would I be looking for? For me it would be about the camaraderie. I'd want to have time to talk with everybody on the tour. I mean, the people are what it's about when you get around these cars. When you get all done, you get car guys together and there's always something to talk about. There's always something to be learned. It's like a fraternity; there's a held interest there that everybody has. It's a gel that can't be found in the general public, right?

I've heard that if you break down on the Power Tour, you'll be swarmed by people with tools and parts wanting to help. And those road trips are done by some of the best rods in the world. If you are brave enough to put it on the Power Tour, it probably is a substantial car to begin with. So you're not the run-of-the-mill hot rodders out there. You are meeting some people who have been at this for a while.

What do we learn from all of this? What are the lessons? I don't know exactly what to say about that. It is a love you carry through your life. I've probably learned more about this and about other businesses because of this, but I don't know if there is an exact lesson I have to tell somebody. Maybe, if it's in you, don't be afraid to play with it when people tell you you're crazy. Because you're not. I looked like an idiot for 20 years, and now everybody is wondering how I did this. It's just like everything else in life; if you learn it and do a good job at it, it's a good thing. You are probably going to make money. Anything you love is worth following and that's where it came from for me. It was just something in my core. I can't explain it, but I am happy when I am playing with it. And the rest of it just comes naturally.

These days I don't do much actual work on cars because I am fortunate enough to have two excellent mechanics. Shy of adjusting the carburetors or something like that once in a while, I have the guys that I am paying to be at the museum shop. I still do have the urge to build a rod, maybe when I am all done and I am retired. But right now I've got several businesses, I've got a lot happening all the time, and I don't have the time to stop and enjoy it. But I could see building one more rod that I'd build 100 percent myself. I've got some ideas and I have the luxury now. I could take the project car in and go put a 5-speed in it, go back and drive it, fix the rear end, and so on. I'd build a whole rod by test and tune—you know, by just playing with it. But I don't actually do wrench-turning right now.

If I built a car what would it be? Well, my first car was a '55 Chevrolet, and I keep going back. I am building a '55 Chevrolet rod right now. I say, "I am building." Well, I am designing it and deciding the engineering on it, and then I have my guys do the work. I just like the '55 Chevrolet. I like the room of it. I like the power for it. It's a solid enough platform; you can overpower it until hell won't have it, and it will take it. And to me, it's a perfect platform to build a hot rod on. I do love the old '30s cars. The '40s—I kinda skipped that generation—the '40s don't do a lot for me. I like them, but not enough that I am going to put a ton of money into them. Right now, my '55 that's on the museum floor, I'll drive it for two weeks and I'll figure what else it needs and put it back in the shop for two weeks of work. I won't let it become a restoration project, where it's gone for two years. That happens, and it's hard to hold your interest.

I had the '55 painted when I first got it—that gets back to my obsession that I can't drive something with primer on it. It's got to look good. But I know that car will be torn down again sometime. The car is beautiful to most people, but it's not arrow straight; it's still a work in progress. There's a dent here and there. The average person looks at the car and thinks he could enter it in a show and win a trophy. But it's not what it should be, and I know that. So when we get the mechanics exactly right, we'll go back and redo the body work. It's more fun when you do it as a project like that. But I can't say it's the most cost-effective way to build a car. It's smarter to pull it to a frame and build it right. But there's an enjoyment in adding this, playing with it, adding that. Right now I am running an LS6 in the car; a 454 big block, a Tremec 5 speed, and a 12-bolt Chevrolet 3.42 rear end. It gives it great power, and with the 5-speed, you can kick it up and drive at interstate speeds all day without heating it up. And I've got air conditioning; it's Florida.

One of the first lessons that I learned with the first hot rods that I built was not to put too much motor in them. Everybody does that. The other thing I learned is if you are building a hot rod, decide on your motor last. Technology changes. If you buy a motor and then build the rod, it's three years later and there's newer motors out there. So you are building yesterday's hot rod all of a sudden. If I was starting to build a hot rod from scratch, the last thing that I'd consider is the motor. And I don't like overpowering anymore, because the truth is, you want to drive a hot rod from town to town. And you want to enjoy it on the road. And you don't really care how fast it is in the quarter mile. A young kid thinks he does. But when you get all done, you can still light up the tires if you want to. You can overbuild a motor real easy for a hot rod.

I've got several Corvette projects with the LS motor and drive train in them. It's warrantied from Chevrolet; you buy it, you put it in, it runs perfectly, it's light, and it fits. It's made it so easy for the hot rodder to build something nice now. Some people say, "Well, it took away me building my own motor" and things like that. Well, things change. The technology is green. It's as clean as any other car on the road when you use these drive trains and it runs. You got a car that's all but warrantied when you put it out there.

You know, I get up every morning and I can walk into the museum and I am looking at a dream. I don't think that feeling will ever go away. I've been around these cars, like I said, for 40 years. And it's the one place I can go every day. Back when I was heavy into construction—which I will be again when it comes back—it's the one place I'd end my day, every day. I don't care how bad a day you had, you can't stand in the middle of that collection and not end up with a smile on your face.

Everybody walking around is happy, and it just brings it out in you, you know? There's a charm to the place. And many of the people in the building are Europeans. It's a cheap vacation for them, and it's a magnet, too. Those people come when they learn it's here; they'll plan their vacation around seeing this place. If anything amazed me about this, it's how far away people will come from. It's kinda rewarding. I did not really know whether it would or wouldn't make it; it was just a dream that I always had. People will come from unbelievable distances to see these cars.

There is a junk collector thing in every car collector. I have a parts department that would rival some Chevrolet dealerships, you know? Over the years I probably have 40 Muncie 4-speeds in a room and I probably got that many other transmissions. I've got tri-power off '67 Corvettes that I've bought over 40 years. I was buying them back when they were $800 and I thought they were a lot of money. Now they are 6 or 7 thousand dollars, and I just pile them up. I won't sell them. If I buy a car, and I change out a part, I keep the part. It takes up mega-room, and it's crazy. You've gotta be a little off-base, I think, to keep doing it, but it's like anything; you are not hurting anybody so long as you are spending honest money. It's what you like and you make yourself happy. That's what's important in life.

You know, in business you don't win them all; you can either lay there and lick your wounds, or you can try another direction. If you worry too much about the negative, you are going backwards. It's hard to see yourself doing it, but you do. I am not telling anybody it's a way to get through life, but my cars have always saved me. When the economy went down and everybody else was having a hard time, I could sell one or two cars and put myself back on my next path. You know, I lost a lot of cars that I didn't want to sell. When times are like that, the ones you want to sell the least are the ones everybody wants to buy. And you let them go.

A long time ago I lost a 396 Corvette that I dearly loved. I was a trim carpenter at the time and I wanted to open a trailer business, but I didn't have the funds to do it. So my banker at the time, he says, "If you sell me that car, I'll make the loan." He ended up with my 396 Corvette. I took the loan from him and that helped me build my first commercial building, which I had full of cars practically before I got it built. The cars have always have been a savior for me. They don't eat when times are bad, you can lock them in, disconnect the batteries, and drain the gas out of them. You can go away a couple years if you need to, come back and they are still all right.

Everything that you buy correctly increases in value, usually. And I have learned a lot about buying cars correctly. Generally you look for matching numbers, but that depends on what car it is. Matching numbers is a lot more important in a Corvette than a Chevelle. You can sell a Chevelle, a Nova, or a Camaro as a hot rod for pretty good money. With a Corvette, you get hurt terribly when you've got the wrong drive train. It really hurts the value of the car. But the muscle cars, not so much, because if you were around the muscle car era, you know if your buddy had a Z28 and you had an SS Camaro, and he wrecked it, you'd stay up all night and put that motor in your car. And so the muscle car people are tolerant of numbers not being right. That's not to say a matching number car isn't worth more, but not big amounts more. If everything else is correct in the car, they are forgiving about the motor.

Other than matching numbers, you need to pay attention to the interior. Dashboards are a big deal; they are hard to make right. Nine times out of ten, if you look at the dash in the car and it's all jumbled up and a mess, then you're gonna build a jumbled-up mess. If somebody takes care of their interior, it can be worn real bad and things like that, but not trashed. There's a difference. And with the body, if the body has been Bondoed up six times, turn around and run. You don't need that. There's too many good cars out there. They don't all have to shine. You can make a lot of money doing an interior and paint and putting them back out on the road. I did that for a lot of years. You can make a lot of money doing that. But you have to know you are buying a good solid body. The first place I go is underneath a car. I lay down under every car I ever picked up.

You gotta know what you are doing when you start buying the early Corvettes. You can buy two cars at once real easy and things like that. All of us that have been around have done it at one time or another. When you are out there, you must be careful; cars get built from the ground up. If you are going to buy one, you should look at it from the ground up. Because the underpinnings are everything. Usually, if somebody takes care of the inside of their car, and they keep their engine compartment clean, and the body looks real solid, it probably is. And you are probably on the right track. But there's a lot more to buying right.

Sometimes you just get lucky. My best barn find happened when I went to buy a dog over in Avon Park. I drove by a barn and I saw a little red Nova. Whenever I drove by I'd see the nose of it; it was one of those things. I had bought a dual personality shepherd and I had to go by several times, and I kept looking at it. One day it got the best of me, and I asked if I could buy it, and the guy said, "Well yeah, I haven't done anything to it in a while. I'd consider selling it." And it was after that I realized it was an L79. I told him I'd buy the car because the money was cheap, even before I got a chance to look at it that hard. It ended up it was a 1966, it was an L79 car. I bought it for, I want to say, $1,400 or $2,000, something like that. When I got all done fixing the car, I sold it for somewhere in the neighborhood of $27,000. Now that was a lot of years ago; today, those cars probably bring $50,000 if they are in good shape. But you don't run into that stuff very often. This car was solid as could be. Why it got parked in this barn I am not sure.

But this car thing is not about the money. If I could put a message in a bottle for my grandchildren, it would tell them that if you have a love in your young life, follow it. You are going to do better in something you believe in and you love. Don't go just find a job to make money; money comes if you are doing something you love. That's what it comes down to. You are happy when you are doing what you like, whether there's money in it or not. And that, in the end, is what matters.

Life is good. I don't think I've ever been happier. I am a workaholic; my wife still asks me when I am going to retire. I am retired as far as I am concerned. I have no vision of doing anything much different than what I am doing right now. I like the construction, but I am a developer. I am not out there pounding nails every day. I am not killing myself. I am thinking all the time to keep my mind active. I have a building full of classic muscle cars. And I am doing just what I want to be doing.

15

Rod Petty (Moab, UT)

My friend Roger Jetter told me that I should talk with Rod Petty in Moab, Utah. He said that Rod grew up in Southern California and was a true American Graffiti–era hot rodder. So I drove out to Moab to interview him. My visit with Rod and his wife JoAnn was one of the most memorable of my odyssey.

ROD PETTY: I grew up in a Southern California town called El Monte. It was well known for rock and roll shows at the Legion. And it was pretty well situated in the center of all the hot rodding that was going on in the LA area. What was so great about growing up there was that on every block there was some guy working on a hot rod. You could go over and they would let you clean up and learn how to sand a little bit, and they knew they were planting the bug in you.

They were guys that were back from the war, they were guys that were racing "roundy-round" tracks and just deep into drag stripping and road racing. They were guys that did not understand, "No, you can't do that." They were guys who said, "Well, I am going to try." It was an era when people were grinding their own cranks in their garage just to see, "Well, if I change that, what would…?" And they were making adapters, to fit this transmission to this engine, or this intake to fit that carburetor, and it was just amazing. That's one of the greatest things that I like about this hobby or sport: We don't let people tell you, "You can't do it."

I grew up with a background in wood. My uncles, my dad, we all had sash and door mills. We made everything out of wood. And then I found out that I could make patterns out of wood, and then I'd just transfer it over. And that was just great. You can weld on it.

It was just a timing thing, I think. Hot rodding was big in the '50s to the mid-'60s, you know. Later, by the late '60s and the Vietnam War, it was different. Like the Beatles changed music. After that, many people maybe lost a little of their interest in hot rods. The war cut back on a lot of it. I know that I get a kick out of things that happened from 1965 to 1968, because I was over in Nam for three tours. I'll get an old magazine and I'll look at the car ads from that period and go, you know, "I've never seen that." It's such a weird thing, because I've lost that time. When you were over there, you did not think about anything except getting home. But when I got back, I really wanted to get back into cars. I just loved cars. I don't even want to think about how many cars I have had since then.

15. Rod Petty

My first car was a '54 Merc, the first year of the overhead valve engine block. I got it and within six months, I went overseas. So my younger brother got it and got into a wreck with it. He got hit by a 1964 Ford XL, the most expensive car that you could buy.[1] And all he did was park in this lane and this gal make a right-hand turn and caught the bumper of that Merc. It ripped that Ford Galaxy all the way down. I can remember the Merc had a little bit of dent. You are not going to bend quarter-inch steel.

I had a series of cars. I had '56 Chevys, I had '55 and '56 Fords, and I had a beautiful '57 Chevy wagon. It was the most expensive thing that I've ever seen in my entire life. It had everything. It was a four-door. I got it from a lady whose husband died, and she did not drive. It was that beautiful silver color that they had. It had all the best moldings and all the best options. Now that I think about it, I don't remember what happened to that car. These days I am seeing more and more station wagons at the shows, I guess because they are affordable and it's neat old metal. Those '57s have nice body lines. I see them lowered. Sometimes that's all you have to do. Just change the stance. And then get back away from it, and you go, "I really don't have to do a whole lot more, do I?"

How did I settle on the look for my '37 Ford? That is basically the way they were when I was growing up. At that time, there weren't as many '37s. Lots of '40 Fords and '39s, stuff like that. But it had to have chrome wheels, because in the period that I grew up in, chrome wheels became the norm. You had to take your old rims down to a chrome shop. If they had a matching pattern, they'd sell you a set and keep your old rims. You didn't just go into a store and buy a set of new wheels. With my car I said, "OK, it's got to have chrome wheels, it's got to have baby moons, it's got to have bigger tires at the back than in the front, and I got to drop the nose a little bit." And I always liked flatheads, so I started working my way through old flathead blocks until I found a good block. I like to use late model blocks, '49-'53, unless it's for a restoration.

The later blocks are better made. I don't use their front-loaded heads, because the center-ported heads for the water work better. They really do. All you have to do is put a ⅜ inch plug in one part of the block. It allows you to use the other heads and be more nostalgic looking, with it in the center. So you go for that, you use a better water pump, and make sure that everything works. Beyond the engine, I use the original seats and stuff. It has basically the original suspension. It's how they were done back in the day.

The paint job is a very traditional rod look. My car has gained a lot of popularity in places I go because it's still the way it should be. The older guys look at and say, "Yeah, that's exactly the way it was." Now they call it old school or whatever. But to me it's just being truthful; I can't build the new stuff. A while ago I built my wife, JoAnn, a '29 Ford roadster. It had a flathead, two deuces, and things like that, and it was a traditional hot rod. That's what we called it. Today some people are calling them rat rods. You know, Offenhauser[2] made the statement about how to tell a hot rod from a street rod. He said, "A hot rod has got a flathead under the hood and a tool box in the trunk. But a street rod has got a Chevy under the hood and a can of wax in the trunk." I always thought that was really true and funny.

Does a true hot rod have to have a flathead? Well, it should, but it doesn't have to. Because back in the day, they ran some pretty high-speed Chevy and Ford sixes that were really good. For me, a hot rod is one that is true to a certain period. Nowadays it's called a nostalgic look. But a hot rod is not perfect. They never were. You just build them to go. A hot rod is made to be driven. A hot rod that just sits in a guy's garage? I feel sorry for him, you know? You get out there and go. My '37 is a hot rod. I've blown tires on the '37 and I've ripped the rear fender nearly in half; you get out there and you fix it and you put it back on. You go from that.

I know that some people think a hot rod has open wheels. Well, that was done mainly on the Lakes, on Muroc and on Bonneville. That was for speed and for looks. My '31 roadster is built that way. That car would not look good with fenders. I'd take that car to Bonneville. I'd have to change the rear, the rear pumpkin, but that's not a problem. Yep, I'd love to take that car out there; it's got a roll bar and it's all set up. I have several gear ratios that would work well there. And if things work out the way I want, that's kind of the goal.

What exactly is a street machine? Well, a street machine is not the same as what I consider a street rod. A street rod still carries a certain amount of "time frame." It has to be pretty much underneath the year of '49. There again, there's all kinds of stuff and engine swaps that goes on there. A street machine could be a clean, nice looking '55 Ford or Chevy. But even now you see some of the Chrysler, Oldsmobile, and Pontiac street machines. I think it's because you can get into that cheaper because they are not Tri-5 Chevys.

All of a sudden now there are other cars out there. And a nice looking street machine has a good looking paint job on it, nice interior, and has some period wheels. Now, I know this may not be politically correct, but 21" wheels on a '55 Chevy going down the

Rod Petty with his 1937 Ford street rod in Moab, Utah (July 2010).

road will not look right. But you put on a nice set of Cragar rims and that sucker screams, "I am a street machine." It just does. And you can go all the way into that same era right to about the '70s. I am thinking about the Chevelles, Camaros, Chrysler products, and the Mustangs. They looked great and you were able to get a nice looking machine from the factory. All you had to do was save up enough money to make the down payment and figure it out from there.

The Shelby Mustang GTs were real street machines. Hertz used to rent them out. People figured out they could rent them, race them, and then return them. They'd bring it back, and the rental people would check the mileage and say, "Boy, you didn't go very far." And they'd driven the hell out of it. Those were nice machines. From the street machines, you branched off to muscle cars. For a while that worked; the cars had real muscle. And every year they'd add a little more. Then we had '73, when the gas crunch really hit, and they slapped on all the emission controls. And then they gave you a six cylinder. That didn't do anything for the hobby. Hot rodding suffered from that.

These days, a lot of us baby boomers are retired. And we are finally in a place where we can put a car together. But I go to shows and I see a lot of gray hair. I got to wonder who will pass on the torch. One of the things that I have seen is that when a young kid comes up and really shows interest, the kid is going to get overwhelmed by all the stuff he is going to get told. We tell him how to do this and stuff, because we are so excited about our hobby. We want to pass on that information.

When you see a kid open his eyes as big as saucers, and with a smile he cannot control, and he just stands there looking at a car? If you are the dude who owns it, sitting in the lawn chair behind it, you get up and ask the kid if he'd like to sit in the front seat. I got news for you. That kid will never forget that. That may be a new ambassador. I tell people we need to have as many folks as we can on the positive end of this because there's always people who want to take our cars away, or tell us we can't drive them more than so many miles on the road. Man, that would be horrible.

You know, we kind of police ourselves. My blown flathead over there passes all emissions. And there are no emission controls on it at all. It's not just about looks; it's about performance. We tune our engines and we take care of them. Here at the town park, they allow us to park on the grass. Most shows you can't do that. If a guy pulls into the park and knows he's leaking a little, maybe waiting to do an engine rebuild, or whatever, out comes a big piece of cardboard. He slides it under his car. Why does he do that? He doesn't want to do anything to hurt that grass. This is the 18th year we have parked on the grass. And all the guys at the show take care of the place very well. Our cleanup crew has very little to do. These guys are a conservation group in itself. That's the way I look at it.

As I think about it, there have been some people that have been very influential in my life, like Tex Smith, who wrote about what a part in a junk yard will fit and what it will work with. That's not really good for the magazine that has to sell advertising, but it's truthful. And Gene Scott, in California. I remember when I was a kid, all he did was flatheads. And I always wanted to work there. Right in Rosemead, California. And he says, "I don't hire anybody unless they are 50 years old." But by the same token, he'd help you any way he could.

If you were a young kid and you really wanted to make a hot rod, there were people running around out there who would do it. I think there's still guys out there who'll do that. And the thing that I see now is younger guys with the new Camaro, Mustang, or the Challenger. At our April show last year, I saw more of those new ones with young guys in them than I've ever seen anywhere. That's great. I don't know where they get the money to buy them, but they can be instantly recognized as in the "car thing." It's, "Hey look, I got this muscle car, I am into this thing."

Driving a hot rod, street rod, or muscle car is what it is really about. JoAnn and I have taken the '37 down Route 66 several times. You just go. I shouldn't tell you this, but I've taken some small chunks of the asphalt, where they've gone with another route, and put it in my trunk. It has no real value to anyone, but I put it on display in my car room. I tell my friends, "See that piece of blacktop? That's Route 66." And I've given pieces of that to car guys that I know. Other people wouldn't understand that. It's amazing that some guys are so excited by a chunk of a road.

When I was growing up, we used to go cruise Route 66 all the time down near San Bernardino or Santa Monica. But since I am here in Utah, I really like going almost to the border of California. We've done that a few times. I like the little town of Williams a lot. It's such a nice little town; it's on the Grand Canyon. And it has my other addiction, which is steam trains. I'll go anywhere to ride on a steam train. I have gone all the way over past Albuquerque. My plan is, when I put the new motor in my '66 Chevelle, to drive all the way to the Route 66 Museum in Oklahoma. I've never been that far.

I really like to travel with a group of three or four guys on a road trip. It's so fun. You check your car out, and it's gotta be working perfect. No matter what you've done to it, sometimes you are going to have a problem. You really hope it's not you because you plan on teasing whoever it is. If you are traveling by yourself and you have trouble, there's a lot of people will stop for you. Thank God our sport has got a good reputation and that people will stop. I have had people stop for me. And I ask them why they stopped. Most of them have nothing at all involved in cars. And they say, "Well, we saw you stopped. Do you need water? Do you need anything?" And I tell them that I've got it under control, but I tell them, "Thanks a lot." I really, really appreciate that. And nine times out of ten, they got two kids in the back seat and their faces are scrunched against the window, looking. So I give them the ah-ooh-ga horn, and the kids go crazy.

There's something about this that gives you neutrality; people stop to help. It's Americana. Cowboys, hot rods, and rock and roll are uniquely American. I know people who broke down in Nebraska. They took these old scenic roads and they are thinking, "I am never going to get help here." They break down and a farmer on an old tractor stops: "What do you need?" And the guy will take you back to his farm, and he has a welder and all the tools. And the wife will make some food for you and some drinks appear. And all of a sudden you are going, "Whoa." I've even had some police stop and take me back into town just so I could get a tire and then run me back out to my '37.

I'll tell you one of the best things that I've seen. We have a show here in town in April. I love seeing a little gray-haired lady with her gray-haired husband walking arm in arm together and they will stop at a certain car. And she'll look at him and he'll look at her and she gives him an elbow in the side. And you know what? I just made their day or year. I don't care what happens the rest of the day; that was so good. And I am not going to ask them what the jab in the ribs was for.

15. Rod Petty

As far as I am concerned, the greatest reality show for car guys would be to get a big garage, put 12 car guys in it—you can have a ringer—and just start conversation about what they did as kids. There are a lot of guys who'd be going, "I did that, too." You know, I give Roger Jetter,[3] who writes stories about hot rodding in Denver, such a hard time; I tell him, "We did exactly the same things that you did. It does not matter whether we were in Iowa, or anywhere." We did exactly the same things on the West coast. We just called it different. Seems every town either has a rock quarry or a small lake. There's a road that everybody goes to drag race. Everybody did that. You just did it. It's so great to find that I was not the only one doing that stuff.

You ask about some of the sayings we used to have? How about "fat arming"? That was when you rode in your hot rod, hung your arm out the window, and pushed it against the door to make your biceps look all pumped up. Big guns. And "James Deaning." We used to James Dean our doors. You'd reverse the door handle so you couldn't get your jacket caught in it. Dean got his jacket caught in his door handle when he went over that cliff.[4] That's the kind of stuff people did. Because we were kids.

Why do I cruise in town or go cross country in my '37? Well, it does bring back a certain amount of feelings about when you were younger. But I'd be lying to you if I didn't tell you that I get pumped up when someone gives me a thumbs up or they yell out, "Hey, nice ride!" I got to say that for the next block I am as high as I can get. I don't have to drink or do anything. A guy just walks up and says, "Nice ride." And I am thinking, "You can't beat that." I tell people that if you are driving long distances or something, take the time when you are filling up your gas tank to look around. People may not come up to you, but there's all these eyes looking. It's not an ego thing, I don't think. Maybe it's a justification of the work that you've done?

I love getting behind a motor home on a highway because of the big back window. If they've got kids, they are stuck right to the window looking at my car. And our town here has a tremendous amount of tourism and a lot of them are from Europe. And they will stop to look at my car. They are all walking the streets, and when I notice them I'll hit the ah-oo-gah horn. They love it. And I think to myself, "How much is that worth?" And when you honk at the women the guys with them are thinking, "I am with a hot chick." They like that.

Truckers like hot rods cars and trucks. You'll always get a horn or thumbs up from them. And if your car is on the side of the road, you'll get an 18-wheeler to stop for you. And they got radios and they will get you help. I've had that happen. I don't want to go too philosophical on this, but to me that says we still have hope for each other. Maybe just having that car sets a different mindset. They will stop to see if you need help. Why? What is it in their mind that allows this? Maybe because some family member was a gearhead and they've been to car shows. If you can get people to car shows, you are building a constituency. When it comes time to vote, they are going to vote to let me keep my car on the road.

There are so many other elements to hot rodding. We are guarding history. Documents are important. I collect that stuff. You can find things in old cars. Maybe people think it's weird, but when I find an old car out in the desert, I look through them to see if I can find something in that old car or truck. If you do, it's priceless. My '66 Chevelle is an all-original car. The color, the upholstery—I have every piece of paper that came with that car, all the way down to the salesman's business card. I have the maintenance book and when you open it up there is the steel plate that they would run across in a machine to certify that the required work was done. I have all that stuff. I have the bill of sale that they wrote up with the sale price. Turns out that car has every option you could get on that car. It's got the 283 engine, the super package inside, bucket seats, clock, console, but it's an automatic. I couldn't figure it out; this car cost more than a Super Sport. And what I finally figured out is that it was a lady's car; she wanted the comfort, air conditioning, a nice radio, and an economical engine. She wants to drive this all the time. So it's got a 283 with a 2-barrel on it. And it can't be red. Figure that out!

My '66 is the early model. The late model has a different sail panel in the headliner. If you don't tell them whether it's an early or a late, you'll get the wrong headliner. And on the late model ones, the little clothes hook is plastic. The early model ones are still chrome. As they make them, there are differences, just little things like that. You don't know that until you have somebody tell you how to take care of a special car. And it's Butternut Yellow. My wife says she don't understand why people come from all over just to look at your car. Well, it's a color you don't see. Also, they are looking at a car that has never been molested. Nothing has been put on it, nothing has been changed; you are looking at a car that everyone knows is an all-original car. It's a survivor, which is what they like.

Rod and JoAnn Petty's Butternut Yellow '66 Chevelle; an all-original survivor car. Photograph by JoAnn Petty (May 2017, Sierra Vista, Arizona).

15. Rod Petty

Memories are important. This is the car I got my driver's license in. Think about it. That was a rite of passage. I can still, to this day, see myself lying in bed, watching the clock to know at midnight I am 16. Now I am going to go down and get my driver's license. And I was so nervous and excited that I flunked the driving test. It's not that I didn't know how to drive; my instructor just told me, "You were just going too fast. You handled everything—the freeway driving—everything. But you got a lead foot." I was just so nervous. He said, "Come back next week, we'll just do it. You'll be OK."

But I was just devastated. And everybody at school knew you were going to get your driver's license. And you had the weekend all ready. In fact, it was really bad because I had set up a double date with another guy. We used my car, but he drove. I told my date that my buddy's girl was on her first date and she didn't want to sit in the back seat with him. It worked.

Anyway, looking back at my car history, would I want to do it differently? Maybe pieces. But I'd still want to end up right where I am now. I wouldn't trade those times. Driving around in our rod fires up good memories; even more than that, being on the road is a spiritual thing. It is. There's a feeling that comes over you. I don't like the word "nostalgia," but if you are driving your car and go back to the same area where you grew up, you get flashbacks. I start remembering some of the people. I remember some of the funny times we had together. When we got caught out of school when we were supposed to be in school. You forget about all the bad things; it's mostly the good stuff. And you think, "Boy, I am glad that I lived through that because if I hadn't done those things, I wouldn't be here today."

I don't name my cars but I totally feel some connection to the old metal. Like the '37; I got a lot of myself in that car. This is a strange little town. With the uranium boom and all the mining and stuff, people bought a lot of hot rods. And made hot rods. And they are stored all over this town. The people aren't even here now. But I know buildings that are just full. It's amazing. My friend Jesse says, "You will never be able to sell that '37." When I asked him why he said, "Because everybody knows that '37. You are so imbedded into that. You are in the metal of that car."

Back in the day, you went out on a Friday night, maybe go to the football game and then you would cruise around. But Saturday night was when you took your best girl out. That's why we go cruising on Saturdays. My wife understands this is very important still. I am taking my best girl out in my best car. It's from when I was growing up. You cleaned it up because you wanted to be seen with your best girl in the car. And there's nothing to me that ever surpassed that feeling. Of course, we went out and we raised heck, but the best thing was the feeling that you got with your best girl when you pulled into a drive-in movie or food place. And I can do that right now. I can take my wife, like we did last night, to go out and get a hamburger in the hot rod at the burger stand. That feeling does not change.

Cruising was so important. It was all about Saturday night. It was what you did. You had to get all your work done, and then you'd get cleaned up. You'd use a dollar for, in my day, four gallons of gas. Gas was about 21 cents a gallon. Then you'd go pick your date up. You'd meet her dad and, if it was your first date, he'd ask about your car because he would not let her go in a piece of junk. He'd want to know where you were working and so on. And all those things were so important. All of those things were, as I think about it today, very important to the building of the character that you are.

You did some things. But there were certain rules; you didn't like some of them. My dad would say, "You need to be home by such and such. And you don't keep that girl out too late. Just because you are a guy doesn't mean you don't have responsibility." And stuff like that. And I did it all the way till the day I went into the Navy.

Today, my wife and I still work on cars together. My wife is a car girl; every car that I've had, she's always got in there and sanded and so on. She'll tell you she gets the grunt jobs. Like under the car. Scraping a bunch of grease off and painting it. Well, I have a bad back and cannot crawl under a car; it's cheaper than paying someone. But she likes working on cars. She says, "When you are taking a car down, when you are sanding on it, you are thinking about the guy that assembled it, and about who had it." And when you are lucky enough to be the second owner or can find out who the other owners were, it's just like a spirit thing. Like old buildings: Old cars need to be cared for because they can give us an insight into the past and the future, at the same time. How cool is that?

16

George Roetman (Vermillion, SD)

I met George Roetman and a host of other good folks on the 2011 Hot Rod Power Tour, considered by many to be the ultimate gearhead road trip. Participants drove their rides from Coco Beach, Florida, to Detroit, Michigan, over the course of a week. I interviewed George, Allen Pearson, and Rick Petersen at Jimmy's Bar in Warren, Michigan.

GEORGE ROETMAN: My "car guy" story started at a young age, when I saw a 1970 Plymouth Superbird roll up at the corner general store in Carmel, Iowa. Also, my mom had a '57 Chevy four-door that she eventually traded for a '61 Ford Galaxie. The significance of a '57 Chevy didn't set in at that time. Late in 1970, we moved to Vermillion, South Dakota, where I live now.

We moved to West Main Street around when I started third grade. West Main in Vermillion during the '70s was a happening place. There was a bend in the street near our house, a great place to pitch the cars a little sideways while coming around the curve. Many Chevelles, GTOs, Mustangs, Darts, and cars like that cruised that street. I had a front-row seat to cruising and a great variety of cool stuff to dream about.

Across the street was a small independent garage called Leo's Repair. It was here that I got started in car repair; by the time I was in fifth grade I was very involved in working around the shop. Leo was an old guy and the shop had two stalls with enough room to work on one side. There were many hidden trinkets (mostly Ford parts) in the shop. The office, if you could call it that, had 428 Police Interceptor engine parts as well as an original aluminum intake. The cabinets were filled with stuff that would be considered new old stock (NOS) gold today.

At some point, maybe around 1975, the guys that hung around got involved with stock car racing. Many '64–'69 Chevelles, including Super Sports, fell victim to weekend dirt tracks. Leo was a Ford guy, so we built a '64 Galaxie using a 100,000-mile 390 engine with a new cam and carburetor. Understand, these cars were very crude and didn't win much of anything. We did try hard to build more competitive engines and mostly ended up with spun rod bearings due to oil loss from higher RPMs. The car was big, long, and heavy; it wasn't very competitive, but we did manage a Midseason Championship win.

Leo was a good guy with great values and took many young men under his wing, sharing and mentoring about more than car repair. During the summers, he would pick me up on Thursday mornings and we would go to the Omaha Auto Auction to purchase used cars to sell. He had a set profit on each car, usually bought one car at a time, and often the customer would be along. Leo ended up going to Florida for the winters after some heart issues. One of his apprentices, Al Richardson, took over the shop as a part-time gig that eventually turned into a full-time independent garage where I spent my teenage years.

Looking back, it seems that cars were always on my mind. I took high school auto mechanics, also attended Western Iowa Tech Community College that led to a 22-year-long career at General Motors dealerships. I am now self-employed in a small independent auto repair shop off Main Street much like the one I grew up in.

Early on, I didn't have the resources to build some of the cool stuff I dreamed about. Now I have more opportunities. Our son Justin has gone to body school and we have acquired some fabrication equipment; hopefully, together we can build some of those dreams. Time is a big thing; I need to fix customer cars before getting to my own.

I have a few cars. My '73 Pontiac Grand Am is originally a California car with a 400 engine and TH400 transmission. We added an Indesign fiberglass nose and NACA duct hood.[1] The car is pretty stock with a few add-ons: 18-inch Boss Motorsports wheels, Ram Air Restorations RA IV exhaust manifolds, and duals with GTO split tips. I changed the wheels back to factory rallys for the Power Tour.

This car drives so good I hate to spoil it. I've Long Hauled three Hot Rod Power Tours and drive my car as much as possible, usually from four to six thousand miles each summer. I need to freshen the transmission and I have a short block ready for assembly. I've got an '89 RS Camaro that I hope to get to the drag strip this summer, It's got a 383 small block that we originally built to run E-85 fuel. I've found less aggravation with 110 race gas. We need to rebuild our extra TH400 and then install it, get a driveshaft built, finish the fuel system and do a little wiring. Then we have to see if it's quick enough for a cage or roll bar.

I still have the '92 Richard Petty Edition Pontiac Grand Prix, another mostly stock vehicle that is fun to drive, and have used it for three Long Haul Power Tours. This car is No. 17 of 1,000 produced; originally it was ordered by Pontiac and delivered to Daytona International Speedway for use at the 1992 Daytona 500. Richard Petty autographed the dash at the '97 Nashville Nationals.

A couple years ago I acquired a '56 Pontiac four-door wagon. It's a rust bucket, but the engine starts and runs pretty well, except for the blue smoke. We got the brakes fixed up this fall and I drove it a few blocks. Ridiculous as it may look in its current state, I enjoyed driving it the short distance. This car is far beyond a reasonable restoration candidate. However, with the rat rod deal nowadays, I think that with a few patches it will be a fun car to take out on cruise nights.

Our kids inherited the gearhead gene. My son started racing Junior Dragsters at 8 years old. The first year he went 14.50 and 42 mph racing one-eighth of a mile. He took a beating the first few years, rarely making it past first round, which I believe helped build his character. He had to learn to lose before he started winning some rounds. Eventually he won the 1998 Thunder Valley Dragways Jr. Dragster Championship. We also

George Roetman with his '73 Pontiac Grand Am at Bunyan's Bar and Grill in Vermillion, South Dakota (October 2016, photograph provided by George Roetman).

went to Indianapolis for the NHRA junior championships that year. He raced High School Series during his junior and senior years; he was runner-up the first year and champion the following.

He now has a back-halfed '91 RS Camaro drag car that runs low 10s.[2] We are going to add front-end limiters to it and change front struts to double adjustable and are looking into torque converter selections. He has a Pro Shot Fogger nitrous oxide system and would like to spray it. We've sprayed it before, but we can't keep the front end down. Wheelie bars are the correct answer, but my son doesn't really want them yet. His engine is a 427 small block making 620 hp on the local speed shop's engine dyno. With the combination we have, it's difficult to have a middle of the road, all-purpose torque converter; it's either got to be set for nitrous or not. I think his car will easily run low nines and high eights when sprayed. He did run a 9.54 at 142 mph with a small hit of nitrous, driving through the converter and on the rev limiter at the big end.

Our daughter also raced some in the High School and Sport Compact Division with her daily driven '05 Cavalier. She has won two Sport Compact Championships and one runner-up. So my kids are hot rodders also; it's a family thing, and my wife goes along with it.

Hot rodding can be defined many ways. Hot rods were the original "jalopy." In many ways, today's rat rods are about people making something out of stuff they have around, or to be different or to make a statement. Hot rodding has been a revolution of ideas and creativity since the first guys tinkered on their rides to make them go faster or look better.

There has been a lot of evolution with respect to speed parts, streamlining, custom billet, and paint. Cars have been turned into beautiful works of art.

Does a hot rod have to burn gasoline? Burning fossil fuel and making some good noise may be criteria for being a hot rod, but I really think we need to be open to hybrids—be it electric or alternative fuels. The simple fact is, if it has wheels and a power plant, there will be a hot rodder trying to figure out how to make it faster, or louder, or how to change the appearance to suit a personal goal.

Hot rods come in all shapes and sizes; the spirit of rodding shows up in many different ways. It's difficult to define an exact meaning. But I think that we need all kinds of enthusiasts to keep car culture alive. Some people have the ability to build a car from the ground up; some people only do the part that interests them, leaving other work for hire. Changing wheels for a more appealing look, or exhaust for a sweeter sound, may not exactly make your car a hot rod. But that is a form of hot rodding. A guy, or gal, may know nothing about the location of the oil drain plug, but enjoys driving their car. They are sharing in the experience.

Cars are a universal experience. Most everyone has owned one, driven one, or ridden in one. Everybody has had a car they liked and one that they didn't. There are favorite colors, disappointing features, and everyone has a car story. I always like to hear people talk about their car experiences and the crazy driving habits of some relative.

Wrenching on a hot rod has taught me that to successfully build a car or to do a quality restoration, you need a well thought-out plan and a clear goal. You stay focused and move forward without deviating from your original ideas, unless faced with difficulty that forces change. Then you adjust. The same thing is true for whatever you do in life; I wish I would have figured this out sooner!

During the summer, I organize a small Wednesday evening cruise group named SODAC's. My brother-in-law Rick found a SODAC's car club plaque behind Leo's garage sometime in the '70s. I eventually tracked down Gary Messler, an original member of the club, and learned the club was formed sometime around 1958. Back then, the abbreviation for South Dakota was So. Dak. They changed the K to a C for their name and sent their logo design off to JC Whitney to produce their plaques. I got permission from some original club members to use the name and logo for my cruise group. We have dinner cruises; sometimes we visit shops and garages. We help each other with projects and share collections. It's a

An original SODAC'S Car Club plaque (image provided by George Roetman).

great way to get people together and make new friends, while driving classic cars and hot rods.

We recently revived an almost extinct local car club, the Dakota Classic Cruisers. As a club activity, we put on a small show and shine on the seniors' last day at our local high school. Students and faculty also participate, showing their cars, trucks, and motorcycles. We provide an unassembled High Performance Chevy 350 engine and a team of four or five students assembles the engine with simple-to-follow instructions and coaching. This is not a speed event, rather an opportunity. The students take a little over an hour to fully assemble the engine. We also have the South Dakota Highway Patrol come and visit with the students. This is our opportunity to give back to our community, to share what we do, and to encourage interest in our hobby with next generations.

While some car guys like being in a club, others just like being part of a no-commitment cruise and occasional garage tour group like SODAC's. Either way, we have a great group of friends that enjoys cruising and the camaraderie created by a passion for hot rods and classic cars.

17

Rick Love (San Antonio, TX)

Rick Love is the executive vice president of Vintage Air, a Texas manufacturer that specializes in air conditioning systems and components for hot rodders. I met Rick at the beginning of a 2010 Street Rodder Road Tour that ended at the Goodguys Colorado Nationals. He has turned his hobby into a profession and also works as a volunteer for SEMA (the Specialty Equipment Market Association) to preserve the rights of automotive hobbyists. Hot rodders everywhere know him as a stand-up guy.

RICK LOVE: I don't know if there is a hot rod gene or not, but as long as I can remember my dad was always interested in cars and still has a '56 Cadillac that he bought in 1957. He always repaired and maintained all of our family cars, and he made some extra money working on his friends' cars when I was growing up. I remember spending a lot of time out in the garage with him watching and handing him tools. On the other hand, I have a brother who has absolutely no interest in cars. To him, a car is like a hammer. It's a tool that gets you from one place to another. It's hard to know what lights the initial fire, but I've been interested in cars since I was a kid, from collecting Hot Wheels and building models to later watching a neighbor's son who was into dirt track racing. You know, back then, in the early '70s, race cars we called Modifieds were running on the local dirt tracks. They had '30s coupe or sedan bodies, and were just cool looking cars that sounded great. He worked on one of them, and I kind of started hanging around with him. When I look back now, we cut up some really nice old coupes that would have made great hot rods.

But the race car guys were also hot rodders. They had hot rods that they would drive, and that's where I started getting interested in and passionate about it. As I grew up, I went through the Camaro and muscle car phase in high school like most of my friends. I had a '74 GTO that was a pretty decent car, and we would run around town and occasionally run it at the local drag strip. I always liked the older cars, though, and then I met an instructor at the local college who had a '39 Chevrolet. I started helping him with it, and I just decided at that point that I really wanted an old hot rod.

We found a '40 Chevy coupe and I got started building it, and that led me to get involved with a great bunch of guys who belonged to Triple Cities Street Rods. From what I've seen, very few guys do cars just by themselves. When you get a group together, it becomes fun to work on other guys' cars because the cars are just part of it. The best friends I have are people in this industry and my hot rod buddies. We all enjoy the cars, but it's the time spent with your friends that have the common interest that are really the best part of this hobby.

I moved to San Antonio in late 1982. At that time, I already knew about Vintage Air from reading all the hot rod magazines. I met Jack Chisenhall at a rod run shortly after I moved, struck up a friendship, and after a time I started working with him on special projects. To make extra money on the side, I started doing AC installations and some street rod wiring. From there, I started doing some R&D work with Jack using the '39 Ford Coupe that I still have today. In the late '80s, when R-134a refrigerant was first coming on the market, Jack had the foresight to see that designing aftermarket systems to work with the new refrigerant was important, so he ended up using my '39 for a lot of the early development work. I probably had the first hot rod in the country with R-134a, and the real-world road testing we did really helped him be at the forefront with Vintage Air.

Over the years, I started doing more work with Jack at events like the Hot Rod Power Tour and Americruise. In 1998, Jack convinced me to come to work for him full time, and I have been here ever since. Even though this industry is my living, I still look at it as a hobby as much as an occupation. I enjoy going to several events a year just as a hot rodder. Obviously, if somebody has a question about Vintage Air, I'm happy to help them, but I go there to enjoy looking at the cars and spending time with friends like I always have.

When you ask me what makes a hot rod in my opinion, that's kind of changed for me over the years. Being in the industry now, I look at it even a little differently because I see all of the changes taking place. If you'd asked me this question 30 years ago, I'd have said a pre–'49 car that had a newer engine and suspension, and improvements that made it faster and more fun to drive. Now the line is a lot more blurred. I sometimes have a hard time looking at what used to be a "late model" car like a '79 Firebird as a "hot rod." That being said, I think the build style can also make a big difference.

Rick Love with his '39 Ford Coupe. In the late '80s, Vintage Air used it for R&D; it was probably the first hot rod in the country to use R-134a refrigerant (July 2013, photograph provided by Rick Love).

The stance, wheels, and color choice go a long way in making the hot rod statement in any build. I have to say I am definitely not as cut and dried about the pre-'49 thing anymore. By any real definition, a hot rod is a vehicle that has been modified for performance increases, so I'm not sure the age of the car matters so much anymore. My '72 Camaro is older now than the '40 Chevy was that was my first hot rod, so that points out a whole different perspective on how old the car should be.

From an industry point of view, I've definitely become more flexible, because the guy that has a '68 Camaro believes he has a hot rod and we need all the hot rodders we can get. He's a member of NHRA, Goodguys Rod & Custom, and even NSRA, which was the last bastion of that pre-'49 rule. They recognized that in order to remain a viable organization—in order to continue to grow—it cannot be just pre-'49 anymore. Do I think the hobby is going to grow? Definitely, and we are seeing more young people get involved these days. Are there a lot of gray beards? Yeah, a whole lot of them; but we are still the biggest generation that grew up in the era where a car was your ticket to freedom.

Things change. Twenty years ago, the lion's share of what we did at Vintage Air was pre-'49 cars, but that's no longer true. We sell far more units for later models than we did back then, and we sell units for cars and trucks up through the late '70s. It's definitely changing. We sponsor NSRA's 29 Below program, and I am really encouraged by the number of younger guys I see. I always hope that some of these kids are going to go from a Camaro to a '32 Ford or the like, but the prices of some of these cars tend to keep the younger ones out of the market even if they have a real interest in them.

If you have been at this for a long time, you realize how the costs of building a car have changed. I probably had $5,500 in my '40 Chevy when I first finished it. It was painted, upholstered, had a pretty strong small block Chevy engine, and was ready to go. Although I did the majority of the work on my '32 by myself and with friends, I guarantee I have a lot more money in it than that! I paid more than that for the body and frame when I bought it in 1995. A 25-year-old kid starting out today may have a hard time getting the kind of money he will need to build or buy a pre '49 car. Some of these young guys go to the tuner cars and then maybe move on to muscle cars and hot rods. From a business standpoint, we certainly hope that's the case. Early trucks and pickups are pretty strong and seem to be getting more popular all the time.

The street rod style build has really morphed itself into the street machines or the muscle cars. You can see it in the shaved marker lights, minor body modifications, and in the engine compartments. That's one of the trends that have really changed in the last 15 years. You're seeing cleaner engine compartments, smoothed firewall and inner fenders, and customized dashes and consoles with different gauges in later model builds. To me, you know, these details can help make a later model more of a hot rod. That goes back to the question, "What makes it a hot rod?" How many changes have you made to the car? If you've just taken a stock '72 Camaro and put a set of wheels on it, it's hard for me to consider it a hot rod. But if you change the sub-frame, install a 4-link suspension, improve the brakes, change the interior and put gauges in it, then you are getting to a hot rod. I have a whole lot easier time seeing that as a hot rod because you are improving the performance and looks at the same time.

In the last several years, we've seen a lot of growth in the truck market. When I say the truck market, I mean everything from the '30s and '40s up through the '50s and '60s, and even into the '70s and '80s. I can look at a pickup as a hot rod. I had an F-100 years ago that was a hot rod to me. But a few years ago, a truck was just a truck; they weren't considered a hot rod to a lot of people. To a majority of us now, that line is blurred, too. You see a lot of popularity in the '67 through '72 C10 pickups. They have always been really hot, along with the '53–'56 F100 Ford. But these trucks have gotten more expensive, so later model trucks are becoming more popular. When I say later models, I mean trucks like the '61–'66 Fords, which are great looking pickups. In fact, we recently released a new SureFit kit, a bolt-in heat/cool/defrost system for those trucks. We thought it would go over pretty well, but it's really exceeded what we initially thought its popularity would be. The growth is even spreading into the square body trucks from the '70s and early '80s. I think some of that has to do with the affordability of those trucks. You can find and buy one for a lot less than a '67–'72.

The same can be said for the '55 Chevy. If you would have told me 20 years ago that builders would be putting a $25,000 chassis under a '55 Chevy, I'd have said you were crazy. But now that they are bringing more money, it makes it more viable and it's a great way to make an old car drive and handle like a new car.

Boyd Coddington was one of the guys that started taking things to another level. I remember looking at his cars initially, and they were amazing. His cars were a little too smooth and in some ways over the top for me, but the workmanship and results were stunning. He really brought car building forward. He was one of the first to do smooth firewalls and flat floors, but they were really expensive cars in their day. I am thankful that there are people out there that support guys like Boyd. They invest in cars like his and push the envelope. I remember, back in the '80s, there were a lot of guys that were envious and real jealous about his stuff. I challenge you to look at one of his cars and not find one thing that is pretty cool that you might be able to use at home on your own car. That's part of what these professional builders bring to the table.

Technology and the aftermarket parts that are available today have helped all of us build cars that are so much better than they were 20 to 25 years ago. I've seen it from the industry standpoint because of the quality of the cars that are out there. You know this if you've been going to shows for a while. Twenty years ago, you could go to a show and there were a handful of cars that really stood out. Nowadays, the average car would have been one of those cars that really stood out 20 years ago. The bar keeps getting raised every year.

I spent 10 years building my '32, and we did a lot of the work in my garage. When I look at the work that Bobby Alloway or Alan Johnson does, there is no comparison. I don't know how those guys could screw a car like that together once the pieces are all done, let alone do that level of work. The level of fabrication on those things is artwork, and it's just cool to look at and appreciate.

But you also have to keep things in perspective. I bought my '39 in 1984 from the second owner in Michigan. The guy that I bought it from put a new motor in it and hot rodded it in the early '70s. We did the bodywork, painted it in my garage. It's a real driver. For a while, it was my only vehicle, and I drove it every day. Years ago, I put a 4-bar and disc brakes on the front, and a 4-link in the back. I did things through the years to improve it and take advantage of some of the newer technology available so I could enjoy it, and I drive the heck out of it. One of the fun parts of playing with old cars for me is making a bunch of parts from other vehicles and different manufacturers work together. I don't have to have a "perfect" car to enjoy it.

As we talk about this, I am finally in the midst of rebuilding the old '39. I took it apart almost two years ago, after putting 180,000 miles on it over the last 30 years. It was at the point where it needed lots of attention, and I decided that since I was planning to keep it forever, it needed to be rebuilt. The motor was about worn out, and the lacquer paint that we had shot in my garage was starting to come off in chunks. So I decided it was really time to go through the car and freshen it up again. Though I really enjoy the process, I don't get as much time to work on it as I used to, but I am finally seeing some progress. We've got paint on the body, the suspension is rebuilt, and the new ZZ4 Small Block and five-speed is installed. I am hoping to have it back on the road next year in time for our Road Tour. Although I seem to be a lot slower than I used to be, I am still enjoying the rebuild process, and I'm upgrading the brakes, steering and just about everything as we go through it. I'm even adding a throttle body fuel injection system to it, just another example of the changing state of car building. We'll be ready for 180,000 more miles!

In our business, we've had to make sure that our products are better and more technologically advanced because people demand it. The other thing that has made a difference is that, if you look back to the '70s when I first started driving my '40 Chevy and look at what a new Corvette or a new Camaro was, they weren't that good a car. They

Rick Love's '39 Ford at the Vintage Air manufacturing plant in San Antonio, Texas (June 2010).

didn't handle well. Now, you buy a new Lexus, or even an average sedan, it will stop on a dime and it's got a really good climate-control system. They are quiet. They are sealed. They are really good cars. If someone is going to put $100,000 in his hot rod, it better be a pretty decent car. It better stop, steer, and drive well. We look at that as a big part of the challenge. We have to make sure that our products are up to the challenge because cars are light years ahead of where they were 15 years ago. And you see that across the board.

What is the state of the industry and the hobby? I think, right now, we are riding the wave of good times in the car guy world. The OEMs (original equipment manufacturers) continue to offer cars with performance that's basically incredible, and that technology is filtering into the aftermarket. The components and parts that are being developed and manufactured in the aftermarket industry are more advanced than they have ever been. The bolt-in suspension options, disc brakes, fuel injection, and climate control systems that are offered now rival the OEM systems in performance and ease of installation. Building, driving, and enjoying the old cars has never been easier or more fun. And I believe these factors are a big reason for the continued growth of the hobby.

The strongest segment of the market right now is in the '60s and '70s muscle cars, and it has been for a while. Part of the reason for this growth is the nostalgic aspect of these cars for the guys who were in high school or a bit younger. They are driving the cars that they had, or maybe wished they had, when they were younger. With the suspension components, brakes, and air conditioning that we have now, you can make them far better drivers than they were back in the day. It makes it a better way to go.

From a bigger picture standpoint, hot rodding has spread globally, particularly in Australia, New Zealand, and Europe. Companies are shipping lots of cars and parts to these countries. Right now, the stronger dollar has made it less lucrative. But builders can still ship a container with a couple of cars in it, pay the duty, do the labor, sell the completed car, and make some money on it. Hopefully, we'll see a continued growth of interest in American cars in some of these other countries.

Here in the U.S., particularly in California, hot rodding is coming under increased environmental scrutiny because the industry has grown. Hot rodders are being held to very high standards. As an industry and as a hobby, we have to deal with this. We are at the point where we are more than a blip on the radar screen.

As healthy as the marketplace is right now, I think it is also so important that everyone, both at the hobby and industry levels, pays attention to the regulation process. And one of the best ways to do that is through the Specialty Equipment Marketing Association (SEMA). From a hobbyist level, SEMA has a division that they call the SEMA Action Network (SAN). This network is for everyone involved in the hobby, keeping rodders informed about both federal and state regulations (both positive and potentially negative).

Several years ago, we were trying to get a street rod registration, titling and license plate bill passed in Texas. I testified before a committee in Austin, and they were very receptive. We were successful in getting a law passed in Texas so drivers now have the ability to register and license reproduction cars and street rods. There are similar SEMA-sponsored bills getting enacted all over the country. We invited our congressman to visit and tour our facilities at Vintage Air to help our representatives understand that although it's a niche industry, we have 140 employees. That's 140 taxpaying families that our little side of the industry supports.

From an environmental issues standpoint, things are also changing. GM is producing eco-friendly engines for hot rodders now. I think we are going to be seeing more of that. Through my volunteer service with SEMA, I have talked with several of the engine building companies, and we've tried to impress on them that we all need to get on board with this. They need to be looking at ecologically friendly engines. It's coming. You are either going to be on the leading edge or you are going to be in trouble. We are already starting to experiment with the other refrigerants that the OEMs are looking at. R134a is still going to be around for a while, but the HFO-1234yf is already being used by several of the OEM companies, and they will all be using it soon. Although it is billed as somewhat of a drop-in refrigerant, it does require some different components to be used. The system we are building for the new Ford GT is a 1234yf system, so we are getting some excellent practical experience with it. The EPA is like the elephant in the room; they have already started to muscle their way in on things and regulations may get tighter over time. I think it is fair to say that the hot rod manufacturers and builders in California are on the front line of the issues.

When you ask where I think this whole thing will be in 20 years, that's a really difficult question. Although it seems like some of the younger millennials don't have the same attachment and passion for the cars and car culture as some of us older guys, the OEMs continue to produce and sell some of the best high performance cars we have ever seen. As long as these cars remain popular, it's a good indication to me that the car gene is still alive and strong. Just look at the number of Challengers and Camaros you see on the street today. With all the talk of the coming development of autonomous and self-driving cars becoming a reality in a few short years, the future is going to be interesting. I am encouraged to see more young people at some of the car shows and events the last few years, and I hope these young people continue to get the same pleasure and enjoyment out of driving as so many of us have through the years. I honestly believe there will always be a segment of our population that loves cars; it's still ingrained in our history and culture, and one of the traits that are just very "American."

The best part of this hobby (and industry) to me is the people that are involved with it. We all share that common love of cars, and there is a real sense of camaraderie. The people that you meet are as passionate about the hobby as you are. I think that's kind of unique to what we do. The best friends I have are people I've either met through cars or from being involved in the industry. For example, John McLeod at Classic Instruments out of Michigan is one of my best buddies. He comes down and goes with us on the Street Rodder Road Tours and we attend a lot of the same events through the year. I met another friend, Kyle Tucker from Detroit Speed and Engineering, the same way—through cars and the industry. Even though these guys live half a country away, we see each other a lot on the road. I'm very fortunate to have good friends all across the country and I can trace them all to hot rods.

My message in a bottle to rodders of the future? It's about the real benefit of our enterprise, whether you are in the industry or a hobbyist. Long before I worked for Vintage Air, I would see the same people every year at the Nationals, and I developed real friendships there. I think that's the best part of this old car hobby. I get to meet a lot of great people that I might not have otherwise. It doesn't get any better than that.

18

Walt Johnson (Thorntown, IN) and Ken Holdaway (Fairview Heights, IL)

I interviewed Ken Holdaway and Walt Johnson at the Kustom Kemps of America (KKOA) Lead Sled Spectacular at Salina, Kansas, in 2013. I overheard them talking at the hotel bar the evening before the show began and I knew that I wanted to talk with them. The next morning at breakfast, I asked if I could join them. I introduced myself, told them about my book project, and they took me under their wings for the duration of the show. What follows is the record of my conversation with them; they were thoughtful and articulate about what they do, why, and about what's ahead as they get closer to aging out of the activity.

KEN HOLDAWAY: How did I get started? My father was a truck driver and in the early years he owned his own truck. He would do his own service; mostly minor repairs. And I would help him, he'd say, "Go get me this or that and hold this or that." That started it a little bit, but it was mostly when I became a teenager. I started subscribing to the little *Rod & Custom* magazine and other hot rod publications. And you start dreaming. And then you got a license and your dad says, "I'll buy your first car." So I got a 1954 Fiat, four door. Don't laugh; I was the coolest guy in high school.

That Fiat was the little and squatty thing. I drove that until probably the last semester of high school. At that time I had a part-time job, and so I told Dad that I wanted to get something different. And he says, "I'll tell you what; I'll give you $500 for the Fiat and you take the $500 and get you a different car." So we went out and looked at cars. The first one I looked at that I really wanted was a 1962 Chevy Impala SS with a 409, dual 4 carburetors, 4-speed and all that stuff. But the wiseness of my father and the cost of insurance for it … well, I didn't get that car. So I finally found a '63 Chevy Impala SS, 4-speed, 327, and ended up with that. That cost me—I think it was $1,800. My payments were like $66 and change for a couple years. I paid for everything including the insurance.

Why that car? I don't know; it was just a neat looking car. It was a 327 car but it had 409 tags on it. And, of course, the guy I bought it from told me this car went to the nationals at Raceway Park in Indianapolis and was a drag car. And they blew the 409 up in it and they put the 327 in it. Of course, at 17 years old, I thought I'd be the most fantastic and hottest guy in town if I had a car that said 409 on it. Which it was not. I drove that car for a couple years. Got out of high school, did a year of college and then went into the Navy. And when I went into the Navy I told Dad, "Well, you might as well sell the car."

So that was kind of the car that got away. Eventually, I graduated from college and started working. After a while when you accumulate enough money, and you get settled, you start thinking about building a ride. So the first car I bought to restore was a '69 Chevrolet. It was just a 383 Powerglide, but it turned out to be a nice car when I got done with it. I had it—after I got done—for three years or something. When I was still in college, I was married, and I bought—it was either a '49 or '50 Ford Business Coupe—just to have something to drive. My wife had the newer, nicer car. And I had that through college and my first job. Then I had a '61 Chevrolet, just a four-door six banger, to get back and forth to work.

But it was my dad that showed me how to do things. He was very mechanical; he did a lot of his own work. Unless it was something major on the truck, then he had to take it to the shop. And when I was in college, my wife's father ran a Marathon gas station and a three bay garage. So I worked a lot of part-time jobs there while I was in college. And, you know, other than changing oil, I learned how to do brakes, minor tune-ups, and put muffler systems on or put a water pump on—things like that. Most folks today don't do any of this. They take them in to get these things done. Well, with cars going to computers too, you are kind of limited to what you can do with the new cars. My granddaughter, you know, I don't think she even knows how to pull the dipstick to see if there's plenty of oil in the car.

WALT JOHNSON: I lived in then what was a rural area up in Washington State. We lived out on the peninsula between Tacoma and Bremerton and so mobility was significant when you were a kid. You didn't want to ride the school bus. And I hung out with four or five guys, all of whom were a year or two older than I was. And they were car guys in a sense that they either had cars or were working to get cars. That was how it was. I bought my first car when I was fifteen, before I got my license. It was a '48 Ford Coupe.

And we were emulating older guys. It wasn't so much out of the magazines. There was a car club there, and I can't even remember the name, but they were about six or seven—and I can still name all the guys in it—and they hung out at a place called Peterson's Chevron Station. I would go there. They built a Topolino drag car, whatever it was, A/Altered or so in those days. And they all had what you would classify as hot rods or customs. I got to be friends with Paul who worked at the Chevron Station and eventually I worked there. On Friday nights often he would take me to a place called Busch's Drive In located in Tacoma. And on Friday and Saturday nights it was like a custom show. They would street race out in front on South Tacoma Avenue. And you would go in and just sit there in awe of the cars and the paint jobs and the power and so on. And so you kind of emulated them. You did things to your car, what you could afford. Top Auto was a speed shop there and you'd pick up a few things here and there, you know.

So that was kind of the start of it. I wrecked the '48 ford within six or eight months after I got my license and I'd already done the interior and leaded the trunk and some things like that. How'd I learn to do that? Well, you taught yourself in those days. The dad of my friend Ronnie Manning was in the surplus business and he had a big shop. Sometimes at night he would let us kids go up there. And he had a stick welder and torches, and so you learned to paddle a little bit of lead and you weren't very good at it but you learned how to do it. And you learned how to put points in your car because your car stopped running one day. And it was almost all self-taught or one of the other kids who had run into the same problem would help you. And it was generally an interchange among a half dozen of us there. But for almost all of us it was just a learning process. For instance, none of really knew how to stick weld but you fooled with it enough until you got it right. And you know, lead was the same way. And I can remember putting a dual carb manifold on and tuning the carbs. And Paul—the guy I was talking about earlier—kind of took me under his wing. He was an older hot rodder at a gas station. He showed me how to fool with the 97s [Stromberg carbs] and things like that. You either learned from somebody older or you taught yourself. You and your friends fooled with it until you got it.

What did these older guys do for a living? They were all mechanically oriented. One of them worked at the gas station, one of them—his dad ran the local newspaper—he took care of the presses and so on. In fact, most of us who got into cars in those days worked at service stations.

They weren't convenience stores, these were real service stations. There were three of them in my area. And I worked at all three of them at one time or another. And they all had lifts and the guys that owned them, once they were satisfied you weren't going to steal their tools or mess up their stuff, let you do more and more stuff. You started up by just pumping gas for them, then they'd let you change oil and maybe they'd let you help them with a brake job and so on and so forth. Eventually—and this is where that hot rod club was at the Chevron Station—he let them use the bays after they closed at ten o'clock at night. And Peterson, the owner, would let them come in at night to work on their Topolino. He wouldn't sponsor them with cash but he sponsored them in the sense that he would let them work on the car. So once I got there and later at a Mobil station, the owners would let you do that. So you had a place and you didn't have a heavy investment in stuff because they had it.

JOHNSON: Was there ever a time when I wasn't messing with cars? Oh, for me there was. I was still fooling around with stuff, for say, probably two years out of high school when I would come home from college in the summer. But then I got married when I was 21 and went back to college. And money is so tight at that point and you can't rationally fool with a car. It was the early '60s, I had kids and I didn't do anything for probably a good 10 or 12 years. And then, interestingly enough, it was custom vans. I had kids but I wanted to fool with the stuff. Not so much the bad airbrush work on the sides, but the interiors. You start out with one with the shag rug interior and you build all this stuff, you put the window in and so on. And—you know—as you look back you think, God, why did I ever do that? But the point is that you were still kind of customizing and fooling with cars and yet it was family friendly. So during the '70s I fooled with that. And then when I got to the '80s, the kids were out; and you didn't need those things and then that's when I got back into it seriously.

HOLDAWAY: As a teenager, in my twenties, I can't remember going to car shows or anything like that. I don't even remember car shows back then. Oh, a bunch of us guys we would cruise where I lived in Terre Haute. The main thoroughfare is called Wabash Avenue, named after the river there, and we would call "Cruising the 'Bash." And there was like an old days a drive-in restaurant on one end, and another half a mile down. And you'd just sort of make the loop; stop at one end and get a soda, and talk with everybody. But it quit when I went into the service. And then I got out of the service, got married, went to college and started working just like everybody else. And you know, when you are trying to get your career started you don't have a bunch of disposable income. It was many years later when I lived in Rolla, Missouri, that I bought that '63 SS and started on it. And I met a bunch of car guys there and got involved in it again.

The thing about my career, and maybe even Walt's, is that our day jobs had no connection to wrenching on a hot rod. You know? It's so much different than in your working life. Sitting in front of a computer or teaching, or whatever you had to do; it's different. Maybe if I was a mechanic then car shows and cars would not be as important as my free time. But I like the process of building a car. It's just something I enjoy doing is working with my hands and that creative process of building something. And once I get it done I get bored with it and move on to the next one. I just like the actual "build" of the car.

JOHNSON: Ken's right, I do enjoy the shows but it's an afterthought to the process. For me, when I was young, it was because that's what the cool guys did. But when I was older and got back into it, I really enjoyed the process of figuring things out. The challenge of building something, of taking it apart, you know. Some guys maybe do model airplanes, some guys may have model railroads; for me it was always cars and mechanical stuff, and trying to figure out how things work. And how to put 'em back together. Like Ken said, that was always—I think—the base of it. And you know, once I got something done and built, I enjoyed driving 'em. We drive our cars. We are not building something to get a trophy. We drive these cars all over the United States. And that's proving that what you did worked. So, as I said at the beginning, a car show is kind of anti-climactic to the process leading to it.

Walt Johnson and Ken Holdaway at a bar near the Sacramento Roadster Show (January 2014, photograph provided by Walt Johnson).

To some extent, as Ken said, when you finish a build you move on to the next project. My wife kind of got me back into it. She knew I'd always missed it and she asked me, "If you could have a car again, what would it be?" Well, I told her that even though I'd had a series of Fords when I was younger and fooling with stuff, that a couple of my friends had '47 and '48 Chevy Fleetlines. You know with the slope backs. I'd always liked the lines of that car. And so she started combing the papers and the Internet and one day she said, "Come on, let's go for a ride." I wasn't retired; I wasn't driving her crazy; she was just being good to me. She just knew it was something that I missed. So she actually found the car. We went down and looked at it. I bought it and I had it 17 years. I sold it last year. I tore that car apart. It took me four years to do the build. I drove that car all over the place. I mean, Ken and I have been back and forth across the county in that car and in one or two that he built. But, as I said earlier, it's kind of the thrill of the build. You want to go back and build something else. And I'd done that car, and I moved on.

HOLDAWAY: Definitely it's the thrill of the build. When I was younger I loved to build models. And when I first started to build 'em I was into World War II planes and tanks and stuff. I kind of got into that. And then when I was a little older, I moved into cars. And that was a neat thing you know, build 'em and paint them up. Then do another; I still got some of them in my garage.

JOHNSON: I built car models too, but it wasn't all cars. I built ships, I built airplanes. So there was that, but another element—I don't think this can be undervalued—is the camaraderie that you develop when you are building the car with other guys. Because there is a community there. And as I said earlier, there is the challenge of learning. When I was in high school I mentioned a couple guys that kind of mentored me and when you get older, that's still there. You learn things and you interchange things and you share information. As old as I am and as much as I know, I am still calling guys. Like I said, it's a community.

HOLDAWAY: Oh yes, I've called Walt about, "You ever experience this?" or I'll call a friend Sam, he's pretty good on electrical stuff, and I'll say, "This is happening and I just don't understand why." I am not really very good at electrical stuff. And he'll give me some hints.

JOHNSON: And there's the H.A.M.B. [The Hokey Ass Message Board]. For a young guy, particularly, because there aren't any gas stations anymore, this is where you can get good information. What concerns me is that I think that I've got a fair amount of knowledge on cars and building these things and I've got a great shop. But there's nobody really to pass it on to. I am 71; my kids aren't interested in it. My grandkids are so busy with everything else. And there's no gas stations left for these kids. Now there is a group of these young kids that are starting to come into this but the concern that I have is that the era when you learned this stuff through these service stations doesn't exist anymore. You may want to pass things on to somebody but it's difficult to find people that are interested.

So maybe that's the role of the H.A.M.B? Absolutely. The website may be sort of a virtual service station. Now it's like any website, you sometimes have to wade through some crap from people that are not car guys or some who are more interested in critiquing or criticizing somebody. But the H.A.M.B for traditional hot rods and customs or maybe the Stovebolt Forum for old trucks, these are places where you can learn tremendous amounts about cars and there are guys that are willing to help you.

HOLDAWAY: I am a member of H.A.M.B. but I really don't use it that much. I don't. When I was building cars in Missouri, I had Sam, Ron, Richie and some of the other guys to help me. If I had an issue, I just asked them. Or somebody else. And back then it was just sort of rebuilding things. So it wasn't like you were "customizing," which as you get into it can generate a lot of problems. Even today, I really don't get on the H.A.M.B. too much. I am too busy playing golf.

JOHNSON: Golf? Christ.

HOLDAWAY: Well, we do go to several car shows every year. Usually we go together to about four and then independently to a couple more. This year we were going to the Stray Cat Customs show in Dewey, Oklahoma, but we had car problems. Almost every

Walt Johnson's shop showing his '33 Dodge coupe, '49 Buick Super Sedanette, and '62 Chevy Brookwood station wagon (May 2017, Thorntown, Indiana. Photograph provided by Walt Johnson)

year we go to Bowling Green, the KKOA in Salina, and the Effingham Frog Follies. We'll go to another one in late September, or October; we haven't decided where. And every winter we fly out to Los Angeles and go to the Grand National Roadster Show.

JOHNSON: We do some shows independent of one another. I went to Minnesota. And I think he goes to cruise-ins almost every week, down in the St. Louis area. And we have local shows. We have stopped going to Goodguys shows and NSRA because they are just too commercial.

You know everybody got upset with NSRA because they opened it to 30 years and older. We kind of understand why because we are getting older and they need fresh blood in the organization to keep going. I don't quite agree with the 30 year thing, but it is what it is. Anyway, the NSRA and Goodguys events cost you a lot of money to take your car in there. And once you get in, if you don't take your own food or whatever, it costs you a lot of money to eat and do things in there. You think about the spectators coming in; they paying a bundle to walk into those things to look at those cars that we pay to get in there. It's just so commercial and the hype is just ... well, it just wasn't any fun for us anymore. So we quit our memberships. And that's why we go to these smaller shows. You go to the big shows you see the same cars or same style of car all the time. At shows like KKOA we find a completely different style of cars and it's usually changing every year. You are starting to see younger people getting involved in traditional hot rodding again. And they are bringing in some really crazy wild looking stuff. It's very innovative. It's just a neat thing if we can keep these young kids focused on traditional hot rodding.

You know, Ken and I are both retired. And we've done well in our lives. So I don't want to come off that it's too expensive for us to be in Goodguys or NSRA; it's not. But as I said earlier, they have become such a commercial outlet. And you know, you go to the shows and there's so much competitiveness for the awards. And some guys are spending $200K, $300K, or $400K on a car to win a foot high trophy And you try and talk to those guys about their car? They have no idea because they didn't build the car. They just opened up the checkbook. And so it has changed over the years from a group of guys who were building their cars or were involved in their cars to guys who are now competing for awards.

Now I don't mean to say that everybody is that way. But the purpose of the thing has shifted. And that's why, over the last four or five years we've shifted away from Goodguys and NSRA to these traditional hot rod shows, where you see people who really are developing and building their own stuff. And really at the forefront; like it might have been like back in the '60s when guys were making their own parts and so on. It's a whole different climate than what you find at the big shows. A young guy really has no way to compete in those shows unless he's sponsored. But there are some guys like Poteet that are extremely good about that.

HOLDAWAY: I don't think we have anything against "builders," because there are certain things that we can't all do on cars. Like our friend Mike Roedar—who you met in the last couple days—he's just a great builder of cars. But I can't afford to have him build a car for me. Plus I don't want him to. I want to do most of the build myself. If there's something I can't do, then I'll hire him or somebody else. You know—more power to him—

he's building cars. But you talk with him and I think you could see his sincerity about the work. Now he's not a big builder that takes them to the shows; he builds cars. Somebody comes in and says, "I want you to do this car for me," but you may never see it at a show as being built by Mike.

If you saw one of Mike's cars at a show, you would say, "Wow!" So he gets work because he does excellent work on a car. It's not because you go into his shop and see a bunch of plaques on the wall or read newspaper articles or anything like that. But Mike is 60 years old now. There are a few younger builders out there, some in their twenties or thirties. Like Dakota Wentz, Darryl Starbird's grandson. He is a kid, but he's the real thing. I am sure he'll make a good name for himself; and more power to him. He's real. He makes a living out of it.

JOHNSON: But as long as we are on that topic, there are a lot of builders out there who will take the money but once the money is there you have a hell of a time getting your car finished. There are a lot of stories about folks taking their car someplace, paying upfront, and never getting the work done. So it gives a bad impression and gets difficult. You have to be very careful about the builders you choose. I know if when somebody comes to me says, "I want to build this car and I want to take it to somebody to help me out with the chassis or this or that." You got to tell them to be very careful about whom you choose because for every ten guys out there, six or seven may never finish the job.

How many times do you hear of guys going to the shop and just taking their car back? Or selling a half finished project because they are fed up with it. And shops open and close all the time. So you must find guys with good reputations. Basically it's a lot like the 60's when a handshake meant that was it. You didn't have to have a ten-page contract. You just shook a guy's hand and he did the work. And finding handshake people anymore is more difficult. But as Ken says, there are some great young builders out there. Guys are doing some really innovative stuff, who like doing it, and whom you can trust. Where I was going with that is that it's a different world than it was 40 or 50 years ago. But that's not unique, I think, to the cars.

HOLDAWAY: What's my next project? I really haven't thought that far ahead. I am still good with the car I have. It's just a primered hot rod and it'll never be finished. I mean, I don't plan to finish the body; other than a few other things that's probably all I'll do for now. It's a '49 Chevrolet two-door sedan with just a standard rebuilt 350 in it, 350/350, 10 bolt rear end, 308 or 311 gears. It goes down the highway pretty good. No cruise, just foot on the pedal; that's about all the cruise I got. I took the heater out 'cause I didn't want to spend the money to buy a new heater core and, of course, no air. It's got power steering and power disc brakes. What am I going to do next? I guess I'd like to put a headliner in it. Maybe insulate it and put a headliner in it. And I probably could upgrade my Mexican blankets to a newer set. You know, to freshen it up.

But I never expected to make it a finished car. She just has a lot of body issues. And as I got older I just didn't want to mess with that stuff anymore. I just hammered out the ones I could and fixed a few other things and shot some primer on it. And that's all this car is going to be. That's all I mean it to be, you know? I bought the car because it was cheap, I thought it was really ugly when I bought it—it looked like a squatty turtle. But

Ken Holdaway's '49 Chevy 2-door sedan (January 2002, Fairview Heights, Illinois. Photograph provided by Ken Holdaway).

I got it cheap. And it was always my "go to" car to work on when I got done with something else I was working on. Like I said, when I get done with a car I get bored, and kind of shove it out of the way after I drive it for a while. So when I get done with one, I bring the '49 in and do some stuff to it until I get another project. So, like I said, it was just kind of a "go-to" car. And then after a while I thought it was just the most beautiful design in the world, which it's not. But there weren't that many of them on the road because people did not like that style of car—the '49s and early '50s Chevy. They thought they looked just butt-ugly. But now you see them everywhere. Also the Olds and the Pontiacs, they are of a very similar design—they are just everywhere in shows now.

JOHNSON: Interestingly enough, when I was growing up in the Pacific Northwest the '49 to '51 Chevys were a fairly desirable car for young people. Then they just kind of lost their aura in the '60s. And it's like they have been rediscovered. But back in those years a lot of young guys that were driving them and hot rodding the six cylinders, you know?

Am I looking for another car for a next project? No, I've got three projects in the shop. Maybe I can finish those. I kind of went the custom "tricked-out" route with my '47 Fleetliner. When I first brought it out—I wasn't winning the Grand National Roadster Show—I got a shelf full of trophies for it. But that wasn't the goal. I built the car as a driver. The longer I was in it, the more I liked just the reliability of the thing. Building something you could drive for a thousand miles and have a hell of a time in and come back—that's satisfying. Things like a completely chromed out engine? That's cool, but over time the trip itself is more important than the car. You focus on what's important— the fellowship. You kind of make a change in your life.

Right now in the shop I've got a '33 Dodge coupe I am about half done with. I've got a '50 Ford that the top is chopped on, and a '49 Buick. Over the last 10 or 12 years I've probably built and got rid of half a dozen different cars; I'll just have to focus on doing these now. Because—you know—at the age I am the window starts to close. And you begin to realize you are not going to have time because a build isn't just a one-year-long process any more. You are taking, depending on the shape of the car, maybe three to five years to build the car. You have to be realistic about things. I am 70 now, and in four years, 74. And it gets harder getting off the cement floor.

HOLDAWAY: I am 66. Well, you know that Walt and I have talked about this; how many more years do you think we are going to be doing this? And especially traveling to car shows. I would assume as long as our health still stays good and we can physically do it that we'll probably do it for quite a few more years. But we've seen that as our friends got older, the number of shows they go to really diminishes. Or they just completely stop

Walt Johnson's '49 Buick Super Sedanette showing its condition when purchased in the early 1990s. He's worked on it off and on over the years, got serious about finishing it two years ago, and expects to complete the build by 2018 (photograph provided by Walt Johnson).

going. And that's just part of growing older and life; we understand that. It's like the World War II vets dying off; now it's the old hot rodders.

JOHNSON: A generation is disappearing. It just is. You go to the shows and you see fewer and fewer of the original customizers—the guys that are in their late 70s and 80s and maybe even older—who made their reputations working on the '50 Mercs and '40 Fords. These were the cars that in my youth were the epitome of a custom. These builders are near or at the point in their life where they are physically not going to build another car. And at some point they are not even going to be coming to these shows. You know you see them, like Winfield, Starbird and others; they may still be pretty spry, but they are not building five or ten cars a year any more.

And then what's ahead? Generally speaking, the 27-year-old male doesn't want to spend $150,000 on a '50 Merc. He wants a new luxury car, so that will be the new hot rod in the future; the newest car with all the gadgets. And all these street rods and hot rods will be sitting out in the fields, just rusting away.

HOLDAWAY: Well, we are all concerned that we have nobody to leave either the cars or our projects to. We just don't. And I often think if I could find a young guy who was interested, I'd be prone to just give him a car. But it's not that I couldn't leave them to family members, some of my grandkids or something, but they—not that they are bad kids—just don't have an appreciation for what it is. And they wouldn't know how to deal with any issues they had with the car. You'd have to leave a long book of instructions. You think of that sometimes and you don't want to stop and sell it off because it's such a part of your life. But you gotta think, "Well … where's it going to go? What's the point?"

JOHNSON: So when is the sell-off going to happen? Well, I've expected it for the last 10 years. We've talked about it. If you look at the age of some of the people you think certainly within the next 10 years. As Ken pointed out, the cars are either going to be sitting in somebody's garage, barn, or shed someplace. Or they will be sold off for next to nothing to somebody [who] doesn't have any appreciation for them. I think the likelihood that they are going to end up in the hands of somebody who is really interested in it is getting smaller all the time.

A lot of people that have big collections are passing away. Maybe in the next five or ten years there will be a lot of cars coming on the market. The best stuff will be picked out by other collectors, but if you are a young guy it might not be a bad time to buy a car, to get some old metal. But what are younger guys interested in? I mean it is generational. Just behind us—four or five years behind Ken, and ten or so behind me—are muscle car guys. You've seen this group; the bulk of the baby boomers. You watch them; they've gone through a motorcycle phase—the Harleys, great for Harley because it's a resurgence for them—and then they got into the muscle cars. But they are not going to move into hot rods. Very few of them are. So there's a big gap there.

And you go back to some of these kids that are in their 20s or 30s and they are like we were back in the day. They can't afford these vehicles. Now there is a small core group of them out there who are in fact traditional hot rodders. They are really into it. But it is a question of how much money they have. They are raising a family. Some of them go the Japanese car route, the tuners. They do some remarkable things with those engines. But there is a lot of money in those things also. And I find them a different group. That's a group who has grown up with more money than you or I did and whose parents have been very giving to them. You know, not to diminish them, but they have had more resources and they are able to do that because they have a financial presence. And can buy the modules that plug in to their cars and some of them do that themselves, but others take it to a shop. So where do they fit in to all this? To me they don't. The tuner guys don't fit into any of this.

HOLDAWAY: I'd say that one generation or segment of people that may carry on traditional hot rodding or customs is the Hispanics in the L.A. area. I think what we've seen the last couple years that we have been out in Los Angeles at the Roadster Show. On Saturday at the show the local people bring their cars and park. And some of these kids are building some magnificent cars and old style customs. It's great stuff.

JOHNSON: You know often times people will label the lowrider group as somebody who has just a set of air shocks who bounces their car along the way. But if you look at the cars, they are well designed, well-engineered and they are running traditional-type engines. Guys are running a '48 Chevrolet with an in-line six in them and you know it's carbureted up and so on. All the original chrome is on it. They have put that great suspension system underneath it; they are traditional rod people. Just in kind of a slightly different direction.

HOLDAWAY: The typical 30 or 35 year old, his ride is going to be the Lexus or BMW. That's his goal. Did we lose that generation to hot rodding? I wouldn't say that we lost it; it's just that cars changed so much in the last 20 years. Since the '80s they were pretty much duds from a design and performance standpoint. And kids carry their electronics—their iPhones—into the car with them so they can email, text, and surf the web. Their focus really isn't on the ride.

JOHNSON: Some of them are programmers and they could trick out their car that way. They could, but I don't know that they are interested in it. For a variety of reasons, as I mentioned earlier; you don't have service stations any more, places where kids can learn things. You have a different expectation level among young people today. They don't even change their own oil. Or do a tune up. They don't know the basics of how to keep a car operational. And—another reason—is that things have become so much more complex. I'd be the first to admit that once we move out of carbureted cars into fuel injection and all the computers and all that other stuff, I couldn't build a car with that. I am still dealing with carbureted engines. Things have become way, way more complex. Now I suppose with some computer knowledge you can learn that but you also have to have some basic knowledge about mechanical and so on. But the two are just not meshing. So for a whole variety of reasons the world has changed; it's not the fault of the young generation.

I look at my own grandkids and I got a couple that probably—one in particular—would spend time in my shop if he wasn't involved in soccer and all the other things.

They are so overscheduled. And sometimes I think they just have no time to be kids; to be inquisitive and to develop things. Their parents, moms in particular, are running them some place every night going from this to that and there's just not a spare hour anyplace.

HOLDAWAY: Probably the only way that the sport will be saved is for dads or grandfathers to have their sons or daughters help them when they work on their cars. That would pass it down through family lines. As far as a new individual coming into the sport, I don't think you are going to see that great an influx because the expectations will not be there. But if you grew up in that atmosphere—going to car shows or working on cars—then maybe it will continue for the younger generation of those families. But for people who weren't involved in it, they just want to get in a car, turn on a car, and go. They are not interested in going to a car show. They just want the car to run, and if it doesn't, they take it to Midas to fix the darn thing.

JOHNSON: Will environmental concerns put the brakes on hot rodding? I read about that possibility but I don't see that as a real upfront danger to the continuation. I think that there are so many other reasons that it's going to continue to diminish that the environmental impacts may just be another straw. I mean, you need people interested in this thing to start with. If you have enough people that are interested, they will fight the laws or the intrusions by regulations. I think that what we've got is a situation where there are just fewer and fewer people interested. And again, that's for a variety of reasons. If you look back, at the real hot rodding in the '40s—maybe the late '30s—it was picked up from the guys coming back from World War II. The gas station guys that I worked for were all World War II vets and that's—some 70 years ago—and you look at the changes that have been made since then, well it's across the board in all kind of things and it's just so much different. Back in the day if you wanted a car you got a job and worked for it. Kids today get a car is as a present for graduation or something. They don't really have to learn anything about it because they just take it down and have it serviced. So it's a whole different climate.

HOLDAWAY: So what message would I like to pass on to future hot rodders, assuming there are any? Hope you enjoy the ride. I don't know; maybe it's just some of these stories we've been telling you. Maybe just how much fun it is to drive what you built and what it's like to experience a great road trip. I'd like them to understand that we built it so we can usually fix it alongside the road. Are we at the tail end of this enterprise? I guess I really don't think about what this sport is going to be like 20 or 30 years down the road. I'll be gone; there's nothing I can do about it anyway. If my cars or parts are still around, I hope that somebody is enjoying them.

JOHNSON: The "message to leave to future rodders" is a tough question. I've heard you mention it over the past two days and I've thought about it. And I don't think I have a really good answer to it, you know? If it was a message in a bottle to a family member or somebody I knew it would certainly be along the lines that Ken said. But a message to somebody that you didn't know who is interested in the hobby 20 years from now? What do you tell them? If you are giving some advice to somebody who wants to hot rod a car, there is probably nothing you can tell them mechanically or design-wise. They are going to do that on their own.

   I suppose the best you can say to them is that make sure that you have some fun. And this isn't any different from cars or anything else that you do. Make sure that you have fun with what you are doing. Otherwise, it's a waste. This is a hobby and it's something you love. What you've got to get out of it—as Ken said—is a memory that you can take with you for whatever life you have. So it's about what a great time you had and how worthwhile it was. Because if it wasn't that for you, what are you doing it for?

19

Jerry Dixey (Austintown, OH)

Jerry is one of my all-time favorite hot rodders. In addition to sharing my love for pedal cars (As a child I satisfied my need for speed by driving my fire engine pedal car down three flights of stairs, crashing at the bottom), Dix is a gifted writer and creative thinker. Jerry is one of the most entertaining car guys that I know. And one of the most thoughtful. As the director of Street Rodder Road Tours, he gets to drive the hell out of a new tour rod every year. I interviewed him in 2010 at the National Street Rod Association (NSRA) Rocky Mountain Street Rod Nationals. We got together again at the 2015 NSRA Northeast Street Rod Nationals.

JERRY DIXEY: I grew up with cars, and my dad had antiques. I have a strong background in cars. I got involved in the hot rod scene in 1989, when I started a company called Classic Transportation. I'd always had a love of collectibles—pedal cars, gas pumps, neon clocks—that sort of thing. At the time, I owned a van and pickup truck conversion and accessory company; part of my showroom was adorned with these "boy toy" collectibles. People started asking me, "Hey, where did you get that?" and I realized there was a car guy market for this stuff. At the same time, reproduction signs and gas pumps were coming onto the market. Not the pedal cars yet.

I realized there was a whole niche business here—hot rod guys that wanted to decorate their garages or basements. Prior to that vintage signs were mostly for antique collectors. I was one of the first to put it all into a full-color catalog: In 1989 I had a 24-page, full-color catalog. I did this clear up to 1995. This allowed people to buy all these things. A majority of them were reproductions, but people could come to my showroom and buy original things. So, when I got involved in the hot rod scene, it was full blown. It was, as I consider it, at its peak from '89 to '91.

You can go back to the NSRA records and see that the numbers really peaked then. I was not around in the early years when they were going to Peoria, Illinois, in the early 1970s when they had 800 cars at their shows. I was a kid then; I read the hot rod magazines, but I didn't go to shows. The changes that I've seen have been on the downslide. It's because of the graying of the hobby. We have studied this circumstance at the magazine and everybody has their opinion, but we are mostly on the same page. What we see is there will be a huge inventory of vehicles that become available. The people that inherit these cars from the dads and grandfathers, some of them—a small proportion—grew up coming to car shows and enjoyed it. They will continue as second- or third-generation car guys. But the majority of people inheriting a car will not keep the car or truck. The increased inventory is going to lower the price of cars.

There's what we call "The Number." This is what you could expect to pay for a good, well-built car that had some miles on it, and you would want to take and make yours by putting a paint job on it and maybe an interior. But other than that, it would be OK. We have seen "The Number" go from $40K, to $35K, to $30K. It's at about $30K now. Granted, that's an arbitrary figure and there is a wide range, but that's where we are at. You can go to Louisville, buy a street rod for that and feel safe driving your family in it. You could turn a few heads for that $30K.

That's about where we sit today. The people who inherit these vehicles are not going to take them to the dump. They are not going to give them away. But they are not going to hold out for the big bucks. They might think, "Granddad had something like $70K in this car. Well, that was probably too much, but that was what he enjoyed. I got it for nothing because I inherited it. I'd like to get $30K for it." That's what we are seeing. As the inventory increases, there is a really big question mark. Are the young kids waiting for the price to come down so they can drive that '34 Ford, or do they want to drive a muscle car? What's going to happen?

Look at the history of antique cars. In 1960, when I was 10, my father and I restored my great grandfather's 1927 Model T Ford station wagon. We still have it in the family. My father and nine other guys started the Model T club of Mahoney County, Ohio. In fact, one of the founding members of that group was Don Snyder senior, who started Snyder's Antique Auto Parts, which is very big in that market. We spent our summers going to antique car shows, picnics, club meetings and that sort of thing. It was a very big deal. But that went away in the '70s and '80s. Guys got old and they died. And nobody wanted to pick up the torch.

If you've ever been in a Model T or an A, you know you really can't go to California in it. Now with the way people want to travel, it's not practical; it's something to play with. It's a Fourth of July parade car. It's not something you get in and drive across the country. This kind of thing is in the mix of what's going to happen. I am 59; the hobby will get me through my retirement.

As to where it's going to be in 20 years? I don't know. There's the problem with fossil fuels. But new technologies are getting into the mix. Factory Five got hooked up with a Silicon Valley company and they are working on putting an electric drivetrain into one of their traditional styled hot rods. Now this may seem kind of funny, but they put a sound system under the hood so they could make sound like a V8. Because they knew that the traditional hot rodders were not going to go for the silent ride. It is all about the rumpety-rump and looking cool. Just like with the Harleys.

We are seeing an influx of Harley Davidson riders who are aging and realize that they can no longer get on their bikes and ride a long distance. Or their wife is no longer comfortable or feels safe as a passenger. As they come to hot rodding, that will be a little bit of a bump. But the $30K number is a street rod number. Since the NSRA and Goodguys have gone up to '72, cars have become a bit more affordable to a young guy, on a budget, trying to pay for a house and raising kids. If you've been around, you regularly hear the old guys saying, "Why don't the young kids get involved in the hobby?" I look at them and go, "Why don't you sell one of them your '32 for $20K so they can afford it?" And they say, "I've got $70K in my car!" So that answers the question, "Why don't young kids get involved in the hobby?" They can't afford it.

The young guys got into imports and tuner cars because you could put a $500 set of wheels on it, add a coffee can muffler, and some graphics and you got the "look at me" thing going. You can't do that with a street rod. With the stretch of years into the '60s and '70s, all of a sudden you can go out to the farm, buy a '52 Chevy pickup truck, drop a crate motor into it, and build it out affordably. By opening up events to "newer" vehicles, we were able to get more people involved.

You ask what the long-term prospects are? You know, I don't know where it's going to go. It's been a good run; they have had 40 good years and seen a lot of growth. I am sure there's other hobbies that parallel this circumstance. You suggested that the old-time tractor and thresher guys might be similar. As you are saying that, I gotta think about how tunnel vision focused people are about their hobbies. I am thinking, "Why in God's name would you want to restore a tractor that you put on a trailer to take somewhere to show it?" But everybody has his or her passion.

Not to digress, but when I was growing up there was just something about internal combustion. It was the ability to get on something and not pedal. And to not get tired. I can remember my first minibike; I built it from the Clinton motor off my dad's lawn mower. I remember going down the road on it and thinking, "I am not pedaling!" Of course, I still have a scar on my leg where the pull-start notch bit me when I laid my leg up against it when I was 10 or 11. The local doc stitched me up on his kitchen table with what looked like fishing wire.

When did I get involved with *Street Rodder* magazine? When I started what became Classic Automobilia, we'd go to shows and set up a diner under a big tent. We had plywood on the ground, black and white checker floors, neon, a counter with stools and booths. It would take us a full day to set up. It was a killer way to display our products. We were just getting into the pedal cars, which were becoming very popular with the hot rod guys. We were coming out with a fiberglass reproduction of a '37 Ford body and a chassis. At the KKOA show in Cincinnati, Tom Vogele dropped by and I recognized him from the magazine. I introduced myself and told him that he should do an article about pedal cars. I said they are getting very popular, people like them and it might be a nice thing to do. He said to me, "Looks like you are the man. Why don't you write the article about pedal cars?"

I'd never done anything like that. I asked him, "What do I have to do?" He asked me, "Do you have a camera and can you write?" I said yes. And he said, "Do some articles." The upshot is that the March 1992 issue of *Street Rodder* magazine has gone down as the "pedal car" issue. I had eight articles: their history, parts available, customizing them, and so on. He liked the way I wrote and took pictures; we got to be very good friends. He went on to ask for an article on gas station collectibles.

At that time, I had a '34 Ford sedan, a stock antique car. I was going to sell it because I was getting into hot rodding. I got hooked up with Barry Lobeck, and he told me that I would get maybe $10K for my sedan. He said, "You are not going to be able to build a hot rod for that amount. Why don't you talk to Tom about an article on your '34 four-door?

Jerry Dixey with a pedal car "treasure" purchased at the 2010 Spring Carlisle Collector Car Swap Meet (April 2010, photograph by Mary Ann Karas).

We'll make it a hot rod." I thought that was a good idea. This was the era of the restomods, and the Pro Street rods, but I wanted to make it a quick and easy hot rod.

So I decided to call it a "presto rod." Tom said, "Wow, that's cool." Barry did the work; got it down and dirty, slammed it, boxed the frame, dropped the front axle, put in a small block Chevy, and touched up the paint. The thrust of the article was that there are a lot of cars out there that you can make into a hot rod without spending $50 to $60K. We did a three-part build of that car at Barry's shop with me shooting it. Got that all done, it ran and the feature ran. In the summer of '94 when it got finished, I drove it for more than 5,000 miles around the Midwest. People recognized the car everywhere.

At the next year's Specialty Equipment Market Association (SEMA) show, I sat down with Tom and I said, "I've got an idea. You know there all these trailer queen guys dragging their cars around. And we've got new guys who are afraid to drive their cars. You know, we never found the skeleton of a dead hot rodder on the road next to a car; they all get 'em home. Why don't we get companies that manufacture parts involved? Let's find a builder, build a car, and take it to all the NSRA events?" Before 1996, no one had ever driven the same car to all the events. Tom said, "We're busy building a magazine. Who is going to do that?" I told him I'd do it; my shop was doing well, and I'd looked at a map and I knew it could be done.

My idea was that everyone would donate parts and at the end of the season we'd sell it so I'd get some money for my time and expenses. Then Tom put an interesting hook on it. He said, "I don't want you to sell it. I think we should give it away. Like a sweepstakes for subscribers." At that point, I am thinking, "OK, but how do I get paid?" then Tom said, "We'll pay you to drive it to all the shows. But I want one number; I don't want gas or hotel receipts. I want one number that you and I agree to for you to drive the car from May to October to 11 events. Figure it out."

I sat down and did the math. Small block Chevy; so I knew the mileage, figured the distance, cost for hotel rooms, and added in 85 days of salary. But, hell, driving around in a hot rod? How much do you really need to get paid? I add all this up and came up with a number. And my wife—who has a couple master's degrees—says, "What did you come up with?" I gave her my number and she said, "Add 30 percent to it." I said, "Why?" And she said, "Because you are forgetting something. We don't know what it is, but you are forgetting something. And a second thing? After four events, after you are done running around and screaming with buddies, it will become a job. Plus, you are going to be gone, you won't be running your business. Add 30 percent to it and give it to him."

You know, I never thought he'd go for my number, but maybe it would be a starting point for negotiation. I sat down with Tom and gave him the figure. And he says, "OK. And we'll probably need to get T-shirts made up and get some other things going." I was stunned. I said, "I didn't ask for enough money, did I?" And he said, "Well, we'll eventually give you a raise." Tom made some phone calls and got my hotel rooms comped for three nights at every venue. Then he told me, "I just got you your raise."

The whole thing has worked out well for NSRA. We made it to all events that first year. And the second year, because '97 was the 25th anniversary of *Street Rodder* magazine, we had to do the McMullen '32 Roadster. I drove that roadster clone 25,000 miles with no top. And the whole effort was very popular. Hot rodders lived vicariously through me as I drove from event to event. Ultimately, I attended 111 consecutive NSRA events. Never missed one in 10 years. At the 10th anniversary, we were able to get all 10 cars together at the Nationals in Louisville. When I saw all them together, it was like seeing old friends. Or maybe girlfriends. There were some that I was happy to see and others that, when it was done, I was happy to get away from. But it was cool; we had a reception. I gave a talk that had me going from car to car telling about each one.

After the 10th year, we had a meeting in LA. And my publisher essentially said, "We need to mix things up. How many times can we go to the Pueblo show? We need you to write an article that invites the readers to drive with you in their cars. It's going to mix it up and will justify why we are continuing to do this." So that's what we did; that was the beginning of Street Rodder Road tours. Of course, a lot of companies lined up to be sponsors; the builds nicely showcased their products. And every year the build changes, so the product mix changes. This all brings us to the present; lots of people really like the tours. About 60 percent of our participants are repeats. We charge $75 to register, but that really doesn't pay our administrative costs. Honestly, that's just to help us know how many participants we are going to have. It'd be a circus if we just said, "OK, just show up and we'll go."

But we keep it manageable, very small compared to the Hot Rod Power Tour. There's

Jerry Dixey at the start of the 2010 Vintage Air Street Rodder tour (June 2010, San Antonio, Texas).

a small-scale feeling to the experience. We don't have the circus at the gas stops. I don't tell people to follow me. I don't think closely following is necessarily safe; we don't want people to run yellow or red lights. We want them to look at their maps and figure out how to get there. But as you can imagine, being responsible for the effort can be very stressful. One thing that we have changed for some of the tours is that instead of driving to a different area every night, we are now staying at a host hotel and taking day trips to local shops and attractions. This is a little easier on some of the graybeards.

But, backing away from the tours, there are environmental issues that rodders have to struggle with. Do I think there will be a move away from the 350/350 to the new stuff? Well, I think the old guys are not going to do it. If they cared about their impact on the environment, they wouldn't be hot rodding. They are reliving their youth, or living out a youth they never had, or they are reliving what they thought the '50s were like—the *Happy Days* or *American Graffiti* thing. And you know, there's nothing wrong with that; it's a family hobby. But you can see the graying, you can see that the people are getting old.

A few years ago, at the 30th anniversary of the East Coast Nationals at York, I shot a group photo of the guys that made the first show. Lots of gray hair and no hair there. But they are still having fun. You go to any small town, and gearheads are having their cruise-ins on Friday nights. In my hometown, it's almost become saturated. A restaurant announces a cruise-in and a whole lot of people show up. Maybe some of the guys are getting stale on that and on some of the indoor car shows. But you know, when I was a kid, I'd only get to see custom cars when a car show came to a nearby city. Now you can go to the Dairy Queen and see really nice cars.

You know, as I reflect on our hobby, we are still at a good time, but I do wonder about how things go forward. There still are lots of hot rodders out there. But we are not recruiting the numbers that we need to make up for the people that age out or pass away every year.

20

Gene Winfield Mojave, CA)

I interviewed Gene Winfield at the 2013 KKOA Show in Salina, Kansas. It is an understatement to say that he is an icon in the custom car industry. At the time we spoke he was in his later '80s; for all intents, he's been building customs all his life. He started his first shop after he returned from Japan after WWII. His shop, "Gene Winfield Rod and Custom," is located in the Mojave Desert. To this day he specializes in fade paint jobs and top chops. As one rodder put it, "The Dude is a legend." And so he is.

GENE WINFIELD: Ever since I was a young boy I liked things that were mechanical, you know? And toys and things like that, I'd try to see how they work. As I got a little older I started building model airplanes and flying them. I built a P-39 airplane; I used wood dough and I filled it, sculptured it and made it beautiful. Then I primed and sanded it and took it to a body shop to have a guy paint it black lacquer. I had two model airplanes that hung in a hobby shop window for several years. Then they closed the hobby shop, took my planes, and I never saw them again. One of them was a Night Twister, a little biplane. It was a beautiful little white plane; I think it was the shortest wingspan of any little plane that ever flew.

So anyway, then I started getting into cars. My first car was a '28 Model A Coupe. And I didn't know what to do with it, you know. I wanted it different. But I didn't know about molding it off or changing it into a hot rod, so just I put things on it. I put on an antenna although I did not have a radio. But I had to have an antenna so I could put foxtail on it. And things like that. On the license plate I had a wig-wag thing that goes back and forth; just little junk that hung on there, see?

Then after I drove it—this was in high school—for a year, I took the head off and I shaved the head. You know I found out that you could hop them up. So then I started seeing other hot rods, and so forth. And then I finally wrecked the coupe; a taxi come out of an alley and nailed me. Wasn't my fault, but anyway, I wrecked the coupe. Then I bought a '30 Model A roadster and then I started really seeing other rods. I went to San Francisco, to Sacramento; I was racing Modesto in Central California, so I started seeing other things. I bought a little six dollar camera and I took hundreds of pictures of everything that moved—anything that had skirts on it or was lowered. I looked for anything that was done to a car that was cool.

Why did I do that? Well, because it was just interesting to me. It was just, I don't know, something in my brain that said, "Man, let's do something that's cool" and in order to do that, "We've gotta take pictures." Those pictures gave me ideas. Have I gone back to those early day pictures? Well, no. Once in a while I look at them, but I don't get hardly any ideas from them because I've already done everything and beyond those pictures, see? But that's how I was doing it growing up.

My second car was a '30 Model A roadster. I put a V8 in it and then some guy said, "You can make your own dual carb manifold by taking some plates and welding them together with tubing." So then I learned how to weld; I got a torch and I brazed this thing together. And I warped the hell out of it and I had to take it to a machine shop to get it machined; to make it flat.

How did I pay for all this? I wasn't rich; hell no. Well, I paid for it little by little. I did

The author, left, with Gene Winfield at the Syracuse Nationals (June 2016, photograph by Eline Haukenes).

a lot of odd jobs, and just anything I could do to make money. So then with the '30 Model A, I started driving it around to other cities and little by little I put a '32 grille on it, I put on solid hood sides. And, like I said, I put a '37 V8 in it and I shaved the heads. I just did all kinds of stuff. And there was a guy named Bill Caldwell that moved to Modesto from L.A. And he was a real hot rodder, an older guy. He's long gone now, but somehow I got acquainted with him. And I went over to his house and he showed me how to do some of this stuff. How to mold off a '32 grille and on and on. I helped him chop a '32 five window in his driveway. Hacksaw and torch was all we had. He was a real mentor.

What advice would I give to young guys getting into the business now? Well, get an idea and stick with it. And even though you don't think that you can do it, that you can't accomplish it, whatever it is, you stick with it and work hard until you make it happen. Don't give up. Work hard, it will pay off. No matter how hard it looks, no matter how unreal it may seem, just do it. Today there are young guns doing this. Like Rich Evans, he's already doing a lot movies and television. He's in Huntington Beach, California. He's doing a lot already. He's got a beautiful shop—Huntington Beach Auto Body. And there's a young kid working with Bill Heinz. I can't think of his name at this minute. But he's doing real well.

What do I think is the future of hot rodding? I've been with it since the early days, the mid-late '40s. I've seen all phases. Some people think the wheels are going to come off in ten years or so. No. The hot rodding is going to continue to grow. And also the custom car field. See, I think the hot rodding has kind of leveled off with all the fiberglass '32s and all the duplicate—you know—cookie-cutter hot rods that are out there. It's leveled off a little bit. But the custom cars are going up. I see more and more young guys getting interested in them. Hundreds of them are chopping cars and all that. And of course, the rat rod—nostalgia rods or whatever you want to call them—that's a very growing thing. What has partly made that popular was *Monster Garage*, the television show. You know they build a car in a week and they show you that they could do just about anything, see?

So that helped a lot of farm boys or home guys that don't have hardly anything, that don't have any money, but they can realize that they can just cut up and weld together anything. And some of the rat rods are just almost totally garbage, and then other ones are just unbelievable engineering. They engineer such fabulous chassis and things to lower them. They are going way beyond what we did. They are supposedly doing what we did back in the day, but they are rusting them on purpose and they don't ever paint them. We had them in primer because we couldn't afford to paint them. And then later we might have painted them. But they are not doing that.

Will this change? Well, eventually it will recycle. Things like that come and go. So yes, that will level off. But right now it's at an all-time high. Will electric cars be hot rodded? Yes, I think that may come. I think that electric cars are definitely in the future. There's no question about it; it's just a matter of time. Of course the biggest problem is the weight of the batteries. You know, if they ever get this solved it's just going to completely boom out there. If they can get that weight down. There are a lot of new materials out there. We are going to see more and more carbon fiber. For sure.

So what is my message in a bottle for hot rodders some 20 years from now? Well, I think I would say that it's been fabulous. It's been a fabulous life, you know? Going through the innovations from my early days to now, through the twentieth century, I think it's just been wonderful and I wouldn't give it up for anything in the world. I think that we've set standards; we've made things happen that didn't seem possible before. And then with the racing; back in the day they said you couldn't go over 150 mph in the quarter, you know? And now we are going 330. And stuff like that; I have seen incredible change.

Part Three

The Run Down

21

The Road Ahead: Higher Gas Prices vs. America's Love Affair with Man Toys

People still have a need for trucks in America and, to a lesser extent, elsewhere. People buy them for work. People want them to haul boats and horse trailers. Not everyone is suddenly going to switch to very small cars, or tiny little pickup trucks, unless they suddenly decide to haul tiny little horse trailers carrying tiny little horses. And there will still be a desire for high-performance vehicles like the Chevrolet Corvette ZR1. There is and will be room for green and mean. Just because a grocery store is expanding its line of organic vegetables doesn't mean it shuts down the meat counter.—Bob Lutz (vice chairman of Global Product Development, GM)[1]

I am going to shift gears for a bit and focus on the big picture: fuel prices and car culture in general. Full disclosure: I am a truck guy; I have always loved classic trucks. For the purposes of bookending my thoughts about hot rodding within the discourse about our current and future transportation landscape, I'll focus initially on America's current love affair with heavy-bumpered, brick-shaped trucks. Then I will pivot to a more specific discussion about the impact of higher gas prices on hot rodding in general.

Consider Lutz's quote as a statement about need versus want. I think he has it partially correct, especially with regard to people who need trucks. However, the marketing for today's full size, Transformer toy-styled, and bull-front-end-styled trucks seems focused mainly on selling them as objects that enhance the manliness of their owner's identity. If you doubt me on this, consider the verbiage accompanying a recent ad for a full-size Nissan: "Tough and menacing," "Super tough," "Bad boy," and "Biggest baddest truck." Even Ford's F-series ads focus on the burly nature of their truck by dropping one suspended in the air or using a front loader to dump a full bucket of debris directly into the bed. Not many non-commercial users would need anything even remotely approaching these capabilities. From my perspective, potential new truck shoppers are being sold the idea that they can increase their manly potency by driving a new Dodge Ram, Ford F150, Nissan Titan, or Chevy Silverado truck.

Hang with me here. On March 20, 2016, I watched on TV the NASCAR Auto Club 400. In addition to male enhancement product commercials, viewers were bombarded with trucks ads. Now, I get the demographic for "supplements," but the focus on trucks during a 400-mile car race strains my understanding. You would think that ads would focus on the brands of cars—say, for what passes as a Chevy—racing around the track. However, almost all the vehicle ads were for trucks. Of course, trucks generate some of the highest profits for car companies. I get that also. But the advertising seems to key on an identity-altering event that accompanies purchasing a new truck.

Granted, many people need a truck for work; folks working construction or doing farm work need carrying capacity. Others "want" a truck to haul boats and horse trailers. So the "My truck is tougher than your truck" message makes sense. I'll also stipulate that fuel economy for full-size trucks has increased in recent years. But when fuel costs rise again—and they inevitably will—the middle class will be squeezed even more and fewer folks will be able to afford big boats and houses. Of course, not everyone is going to "suddenly switch to very small cars, or tiny little pickup trucks ... to haul tiny little horse trailers carrying tiny little horses," as Lutz facetiously argues. But fewer folks will be able to afford the price tag and relatively greater operating costs of a full-size truck. I wonder if Lutz has considered that maybe some folks will have to permanently park their man-toy trucks and start riding their horses to work. But I digress.

How does this all relate to the viability of hot rodding? What are the effects of escalating gas pump charges? I believe that there is a gas price point that will strangle our enterprise. Speculation is difficult; however, consider what happened to non-commercial use purchases of new trucks during the last period of spiking gas prices. When Lutz wrote his *Newsweek* piece in spring 2008, gas prices were near $4 a gallon.[2] Although sales of trucks and SUVs were sinking like a drowning man with rocks in his pockets, Lutz stated that he expected serious change in the U.S. fleet composition after gas passes $10 a gallon. He suggests that until then, the proportion of vehicle types may change, but small cars will not wipe out the large car market. Also, he is confident that there will "still be a desire for high-performance vehicles" and for "green and mean" vehicles.

In hindsight, Lutz's notions seem well off the mark. With gas at $4 a gallon, Ford's popular F-series truck sales fell 33 percent, Dodge Ram sales fell 38 percent, and Chevy Silverado sales dropped 44 percent.[3] Chevy's "like a rock" slogan, also once used to sell Silverados, seems eerily appropriate. Although the bottom did not entirely drop out of the big truck market, it would appear that non-commercial consumers were looking for "smaller horses," i.e., more compact vehicles with high fuel efficiency.

Understandably, it is tough for manufacturers to give up the high profit margin associated with trucks. But when gas prices were going crazy and the bottom was dropping out of the housing market, it seems clear that the green visor guys in GM's corner offices were not convinced that big trucks, SUVs, and Hummers had much of a future. A month after Lutz had his say in *Newsweek*, GM announced that it was closing four of its plants, mostly those making big trucks and SUVs, and that it would probably put the Hummer brand up for sale. They justified these moves with the statement that the market now favors smaller, more fuel-efficient cars. They said they believed this is a permanent change by consumers, given the prospect of ever-increasing fuel costs. My takeaway is that the writing is already on the wall, and not just for trucks. Other elements of car culture may well be impacted at $4 a gallon gasoline.

Lutz may be wrong on the $10 a gallon break point, but is he on target about the

21. The Road Ahead

Tony Farnetti of Sand Hollow, Idaho, with his 1935 Ford pickup truck (July 2010).

future of "mean and green"? Perhaps. More likely, our definition of high-performance vehicles is destined to change. My '58 Apache is green and mean. Beyond her actual paint color, she is re-engineered to get pretty decent mileage (27 mpg)[4] and run clean. She is quick, but she still is a real tank. She is armored in the heavy old metal of a ¾-ton farm truck. No lightweight aluminum, plastics, or exotic composites on her; she carries lots of mass, requiring lots of energy to get moving. I want to keep her on the road, but I'm a realist. At some point (maybe $6 to $8 a gallon?), she's parked and my distance cruising days are over. Some very wealthy folks may be able to keep their rides on the road, or purchase a new Corvette, but the rest of us will be driving small, fuel-efficient and—hopefully—peppy little cars that are green and cute.

It would be a real sea change if manufacturers were forced to produce cars and trucks that are green and "cute," instead of green and mean. Industry ad writers, like many of us, are clearly not ready to deal with such a fundamental change. They are going to have real trouble framing the loss of the big V8. Consider a 2008 Chrysler ad in *Barron's* that asks, "*Who took the wheels off your American Dream?*" It goes on to say

> What happened to the love affair? When did our beloved modern mode of transportation become a need instead of a want? We are still madly in love with our cars. But one visit to the gas pump and you know that love is being tested.[5]

The ad states that Chrysler is working to keep full-size SUVs on the road with a "fuel-efficient hybrid two-mode HEMI powertrain." The fine print at the bottom of the page gives the preliminary EPA estimate of 18 city and 19 highway mpg. For context, when this ad appeared, a barrel of crude was $139. The ad ends like this:

> We live in a car culture. Don't apologize. Without our cars, we lose a measurable source of freedom, individuality and, yes, happiness. That's why we consider it our obligation to improve the technology within the car so your experience behind the wheel remains blissfully unaltered. Suffice to say: we can collectively get where we need to go and still enjoy the ride. Don't garage your American Dream. Put your wheels back on. And drive.

Really? Remaining blissful is a non-starter when car love is crashing on the rocks of high fuel prices. When this ad copy surfaced, Chrysler's best efforts resulted in their trucks getting just 18 or 19 mpg. But high fuel prices consumed any savings that might be gained by modest mpg improvements. Chrysler is doing the equivalent of rearranging the deck chairs on a sinking *Titanic*. Enjoy the ride while you still can; the wheels are coming off. OK. That was pretty harsh. Actually, I really like this ad. We do live in a car culture; cars and trucks are inextricably linked to individual mobility and the freedom of the open road. But there is no subtlety here. The ad suggests that we don't have to drive wimpy cars because Chrysler is working to improve the fuel economy of SUVs and high-performance vehicles. And that we will always enjoy Chrysler rides because performance will not be sacrificed for fuel economy.

This copy skirts dangerously near the "don't worry, be happy" approach to problem solving. To their credit, Chrysler acknowledges that the high cost of fuel is a problem. Perhaps future improvements in fuel economy gained by the more efficient hybrid HEMI powertrain really will be impressive. The ad also notes the company has 11 Flex Fuel capable vehicles. Of course, I'd like to believe that America's love affair with fast cars, powerful trucks, and street rods will remain blissfully unaltered. After all, I built my '58 to run the open road, powered by a fuel-injected and computerized V8 that provides relatively good fuel economy (27 mpg highway) and performance. And some of today's full-size trucks are matching my fuel economy. So maybe it's not all hyperbole, and we can "collectively get where we need to go and still enjoy the ride." That's the hope: fuel economy with no sacrifice in performance.

But hope is not a plan. And magical thinking is not a solution. Keeping the love affair with high performance alive and the wheels on the American Dream will be no small task. Granted that there is a difference between a vehicle driven to work and a truck or hot rod driven for pleasure, but there is no free lunch. Gas costs money. Cruising culture might be parked by a combination of exorbitant gas prices and concerns about environmental impacts.[6] Looking down the road, it seems clear that our definition of "high performance" will change. Hot rodders may come to define performance in terms of extreme fuel efficiency and ultra-low emissions. For this to happen, hot rodders will have to morph their rides by changing to more efficient power plants. Our love affair with big trucks, muscle cars, and hot rods will come to embrace smaller, composite-built, streamlined, plug-in-to-charge, high mileage, and quiet three-wheelers.

Finally, about the opening and closing lines in the Chrysler ad, "*Who took the wheels off your American Dream?*" and "*Don't garage your American Dream. Put the wheels back on. And drive.*" What are we to make of these statements? They surely do not apply to the dreams of American hot rod owners. The ad never says what exactly is responsible for parking our rides, other than inferring high gas prices ended the affair. And the logic behind the instruction to just get back in and drive really makes no sense at all. At some price point, many of us won't drive them because we cannot afford to fuel them.

Updated technology in new cars does nothing to improve the mileage of the hot rod or custom currently in the garage. Apparently, we need to drive our old ride to the local Chrysler dealer and get new wheels. For those of us who love old metal, this is a nonstarter. Perhaps getting where we need to go requires that we exchange our old power trains for new ones. Yet rod builders using these computer controlled engines will have to overcome a good deal of resistance from traditionalists. It's complicated.

I began this chapter with the words "need versus want and desire." Like it or not, high fuel prices and ever-increasing levels of income inequality will change modes of transportation. There will always be a need for commercial transport but the level of "want" for a non-fuel efficient vehicle will decrease. Transportation to and for work is a need; riding in a hot rod is a want and desire. Let's be honest: Hot rods are the quintessential man toy. When gas prices are over five or six dollars a gallon, "wanting" cars that represent a source of freedom and individuality will require additional resources that fewer will be able to muster.[7] So cute becomes cool? How are hot rodders going to ride with that?

Today's muscle cars, like a loaded new Camaro, run well over $60,000. This is a level of discretionary income for fewer and fewer. And given the steep cost of running an old metal, high-performance rod or truck that gets less than 12 mpg on ethanol-free 91 octane, the hot rod in the garage may be driven only on short trips. Today's relatively low gas prices (~$2.43 a gallon in March 2017) may appear to have given our cruising desires a Viagra-like boost, but untreated ethanol-blended gas melts hoses and gaskets in traditional engines. And there is no way that prices at the pump will remain low. So we get a little bump of fun now. Also, over the long term, environmental costs of pump gas will have to be paid no matter how much surplus oil is sloshing around on the world market.

But let's not put hot rods up on blocks or drag them out to the back lot weeds just yet. All is not lost. When I set out on my first cross-country interview effort in 2009, I wondered about the affordability of my enterprise. Gas was more than $3.50 a gallon. My initial road trips were very expensive. There seemed little hope that gas prices would back down. It appeared that the window for long road trips in old metal was closing much faster than I'd ever imagined possible. For an increasing number, road trips such as the Hot Rod Power Tour and long distance cruising in general were on the bubble. And that bubble seemed to be popping. But then, gas prices took a nosedive because of domestic fracking-related production and international events. My early project notions about the imminent doom of hot rodding caused by high fuel costs seem to have been premature.

Nevertheless, before you call bullshit on my concerns about the impact of high fuel prices, I'll say that no one that I know expects that pump gas prices will again settle below the two dollars a gallon cost that we briefly experienced. Hot rodders also have to fill with a higher octane blend, which is more expensive. In addition, consider the even higher price for ethanol-free 91 octane gas, which must be run in older engines. Hot rodders have learned that ethanol gas can give engines the equivalent of an aneurysm as it "melts" the rubber in fuel hoses and gaskets.

And finally, there still is the matter of environmental costs, which have generally been externalized. By this I mean that true costs of production and use of fossil fuels are not being paid.[8] Like it or not, climate change due to fossil fuel consumption is a reality. When rodders run their engines, they are creating pollution. As a libertarian, I've always thought that an individual should be responsible for his own fare. There is no free lunch. We seem to be living in some kind of a grace period where we are not taxed for the full impacts of our hobby. At least for the moment.

22

The Takeaway: Thoughts on the Future of Hot Rodding

> *How different will tomorrow be from today? Both a lot and very little. More of us will live to be 100. Our resources will diminish while our technological capacity grows. Stuff will get faster and cheaper. But our basic needs? We're betting those stay the same—that humans will still need to sleep, to eat, to work, and to move from place to place. That last part is what we're interested in here. What happens to mobility in the next 15 years?—* IDEO Automobility website[1]

Hot rod and custom owners express a great deal of anxiety about who will carry the activity into the future. Most of the guys at Goodguys, Right Coast, and NSRA car shows are graybeards; aging baby boomers. Talk with many of them and they will lament their kids' lack of interest in their rods. Many note that their grandchildren spend all of what passes for free time playing sports, using social media, or gaming online. Gearhead granddads find it almost impossible to derail the inertia created by these activities. Very few kids show any interest in their granddads' Old School hobby.[2]

Most graybeards say if the kids don't get involved, hot rodding will fade away. Nostalgia-based enthusiasm will keep things afloat for only so long. Some observe that younger guys are building really remarkable rides. My impression, however, is that many of these young builders are crafting cars and trucks for deep pocket boomers, with some exceptions, of course. Others point to the development of the West Coast's vibrant rockabilly car culture. The genre includes mostly low- and medium-buck builds and seems genuinely family-friendly. Of course, the Lowrider community of the Southwest is alive and well.[3] And friends tell me that there are a lot of young guys building traditional hot rods in Texas. If there are elements that do persist, these may be among the survivors.

But the deck is stacked against persistence. Future generations may well come to understand this period as a boomer-fueled renaissance. Going forward, things may fall apart. The activity and the industries it supports will face pressure from ever-tightening environmental-based restrictions, affordability related to income inequality, dwindling retirement savings, and high college loan debt many young people carry. And then there is the development of driverless cars. Driverless cars have the potential to do to "analog" vehicles the same thing the car did to the horse and buggy. Old men and old metal may have no place in a country whose transportation landscape is dominated by autonomous vehicles.

If hot rodders and cruisers want to keep the highways open for their rides, they will have to pay attention to the movement toward a driverless future. Groups like the Specialty Equipment Market Association (SEMA) will need to be even more aggressive about protecting the rights of traditional car owners—especially hot rodders. The road ahead will be difficult because of the rapid pace of technological change. New transportation options such as Zipcar and Uber will decrease the number of people who own cars. And it will be hard to throttle back the larger designs of tech companies and venture capitalists:

> The rapid rate of technological change in transportation has been a challenge for the public sector because of the difficulty of keeping up. For better or worse, the government's ability to regulate urban movement has been undermined by the speed of the tech companies and their publicly attractive insistence that they're only increasing mobility. David Plouffe [former Obama adviser hired by Uber] claimed at the 2015 International Transport Forum that Uber is "hungry for new regulations," but it's hard to avoid the sense that Uber simply won't accept regulations that don't fit with its revenue motives.
>
> The rise of tech companies effectively making their own rules and then asking the public to accept them puts into question the government's ability to maintain stability in an industry while ensuring safety and continued access. Is the public sector abandoning its role in favor of crowdsourcing and crowd ratings?[4]

Make no mistake: A process is underway that intends to replace private vehicle ownership with car sharing, driverless rentals, and fleets of self-driving cars. Urban areas, such as gridlocked Los Angeles, may be among the first places where a transition away from car ownership gains traction. Consider the comments by Gabe Klein, with Fontinalis Partners, a venture capital firm that focuses on technology and transportation-related start-ups. Asked what can be done to defeat gridlock, Klein responded

> The single-occupancy car is not good. Do we want to keep buying the cow, when what we really want is the milk? We need to develop a car-light lifestyle. Uber, Lyft, driverless vehicles, robo-taxis are steps in that direction.... It has been estimated that combining self-driving cars, car sharing and various disincentives for car ownership can eliminate up to 85 percent of the automobiles in cities. This would be a sea change for the car industry.

Asked how to get such a change, Klein suggested

> By increasing the cost and inconvenience of owning and operating a car. You can raise parking fees and reduce or eliminate street and off-street parking. You can charge higher vehicle registration fees and higher sales taxes for cars.... Laws also could be passed to limit the number of cars people own.[5]

Car culture has very deep roots in Southern California. So it's not hard to imagine that there would be a good deal of resistance to efforts by well-intentioned environmentalists to push LA gearheads to what Klein calls a car-light lifestyle. Things may not be working from a gridlock or environmental sustainability standpoint, but pushback will be strong from those who value car ownership and the freedom of the road. If you think I'm overstating what's on the horizon, consider a BBC commentary about autonomous vehicles:

> Like it or not, the driverless car is coming. At this year's Consumer Electronics Show in Las Vegas, Mercedes-Benz reported that its new E-class sedan had earned a Nevada Autonomous Driver's License.... And Kia pulled the wraps off an entire sub-brand, called Drive Wise, devoted to driverless cars.

So as we contemplate a driverless future, and examine the ways autonomous vehicles will revise our thinking about everything from the insurance industry to law-enforcement practices, it can be enlightening to consider the end-user changes, as well. Children of the driverless generation will let go of more than just the go-turn-stop skills required to operate a motor vehicle, they will lose many of the attitudes, interactions and cultural norms—the good and the less good—that accompany manual control of a car.[6]

I frequently hear hot rodders lamenting their kids' lack of interest in their hobby. But few are thinking about the effects of driverless cars on the survival of their passion. Consider the cited benefits of steer-free riding: no speeding tickets associated with frisky driving, everyone drives the same way, with no freestyling, less gridlock, better gas mileage, no fender benders, no road rage-inspired bird flips, no roadkill, and no drag racing or burnouts. Many proponents of these changes envision only the upside. Driverless cars will improve safety, cut pollution, and as Sarah Hunter, Google's head of public policy says, "Self-driving cars offer freedom and quality that we never thought of."[7] End of story.

My intention is not to drag you too deep into the weeds with respect to my worries about the future of hot rodding. Instead, I will explore some of the thinking about tomorrow's transportation landscape likely to restrict the activities of hot rodders and highway cruisers. It is clear to me that how we roll in the future will be inextricably connected to the degree to which driverless cars replace our current vehicles.

Planners supporting vehicle robotization seem to assume people just want to ride—not drive—and that they don't want to spend any more time in their vehicles than necessary. Obviously, these planners do not have the car guy gene. Imagine, if you will, the total cluelessness and confusion these folks would experience upon reading the following review of the new MX-5 Miata sports car:

> Probably more fun than the Ferrari, McLaren, and Porsche because you can use it to its max…. A little sweetheart. Just like the original Miata launched more than quarter-century ago, the new edition dispenses joy with its tach needle. Hurl it around the track at the limit, and it rewards with evocative steering, a playful but predictable chassis, and a 155-horse, twin-cam 2.0-liter four that'll gun to 6,800 rpm all day. You'll giggle in the driver's seat—the Miata is that entertaining. It's equally attuned to the open road; you couldn't feel more in touch with the tarmac if you crawled across it on your hands and knees.[8]

There is a takeaway for transportation planners here. Car guys drive: We don't ride. There are fundamental incompatibilities between the proponents of robo-cars and analog drivers. Silicon Valley entrepreneurs and their venture capital backers would improve mobility in metro–America by shunting drivers to an ever-shrinking number of "analog" lanes. Why? Because the driving patterns of individual actors cannot be easily accommodated in computer managed traffic flows. And there is a class element to the new transportation paradigm. The economically more fortunate will enjoy driverless vehicle express lanes. Ultimately, analog drivers will be squeezed off highways and be forced to take whatever passes for "public" transportation.[9]

So how does the specter of driverless cars with priority right-of-way sit with Middle American values? From flyover country perspective, these transportation planners and techies are living in East and West coast bubbles. We've got a real cork versus pull tab thing operating here. The rank and file car guys I know have no interest in constriction of the open road so the top one percent can get to work a half hour earlier.

This is not how we roll. It may be that driverless cars will improve mobility for a fortunate few, but I am willing to bet that other segments of the population will also actively resist being pushed off our highways and limited to riding in a box.[10] It will not be easy to take the wheels off the American dream.

23

Street Rodding 2.0: How We Roll Forward

So [Patrick in Alaska] you would like to take the entire drive train of a Prius and you would want to put that into a Model A? And you would have to put in the wheels from the Prius because you need the regenerative braking and all that.... Is it doable? Of course it is. But, I don't know. There are some things that shouldn't be messed with. On the other hand, how else does humanity make progress without crackpots, screwballs and nut jobs and people willing to do something that's never been done before? Yeah, I think you should go for it. And keep us posted.[1] —Tom and Ray Magliozzi [of *Car Talk*]

Snakes shed their skins in order to grow. A reverse version of that will keep hot rodders rolling for a while. Old metal skins will be stretched over new and alternative drive trains that include computer-controlled traditional engines and hybrids. And from a safety standpoint, technologies such as adaptive cruise control, lane-change assist, blind spot detection, traffic warning messaging and collision avoidance will be crowdsourced for our street rods. After all, rodders are gearheads. And there are massive numbers of car guys connected on the web. Aftermarket kits will be developed to retrofit these products into our rides, and many of these safety features will be readily incorporated into our builds.

Further down the road, rodders will incorporate drivetrains from electric vehicles (EVs) and extended-range EVs. This will effectively mitigate concerns about negative environmental impacts associated with hot rodding. From a performance standpoint, street rodders will give up nothing. Take, for example, the performance of Tesla's new fully electric SUV. *Motor Trend* magazine reports the Model X is the quickest SUV they have ever tested:

> The Tesla Model X P90D in Ludicrous mode ripped its way to 60 mph in just 3.2 seconds and to the end of the quarter mile in 11.7 seconds at 116.0 mph. That beats out many high-performance SUVs including the 2015 BMW X6 M (3.7 sec; 12.1 sec @ 114.3 mph), the 2016 Mercedes-AMG GLE63 Coupe S (3.9 sec; 12.5 sec @ 110.5 mph), and both the 2015 Porsche Cayenne Turbo S and 2015 Porsche Macan Turbo (which posted nearly identical numbers of 4.2 sec; 12.9 sec @ 107.4 mph and 106.2 mph, respectively).[2]

Keep in mind that the tested Tesla SUV has three rows of seating for six occupants. Compare the Tesla's performance with that of the new 455 Camaro SS coupe ("the most powerful

Camaro SS ever") that goes from 0 to 60 mph in 4.0 seconds and completes the quarter-mile in 12.3 seconds when equipped with the eight-speed paddle-shift automatic transmission.[3]

The specter of driverless vehicles and the new business models transforming our transportation system portend a rough ride for street rodders. In the short term, technologies such as lane change and collision avoidance will continue to require full driver engagement. As these technologies grow more common, people will increasingly accept vehicles requiring minimal driver intervention. How vehicle drivers will coexist with self-driven vehicle riders remains to be seen. It seems obvious that planners could minimize the significant technological hurdles associated with blended highway traffic by restricting or eliminating self-driven vehicles. A "no human drivers on the street" policy would ease progress toward the rollout of a self-driving car. So rodders can expect pressure to get their vehicles off the road. The end game of those who support mass adoption of automated vehicles, though they may not say it just yet, is to eliminate the need for private vehicle ownership.

Planners seem to start with the notion that people don't want to spend any more time in their cars than they must. And that the automated car will improve their quality of life, especially in urban areas. I guess they expect car guys to abandon their rides once they experience "car light." This would require future rodders to come up with a new definition for "My ride." I just don't see these guys giving up their steering wheels, lounging on a robocar's couch, and enjoying the ride.[4]

My point is that the impacts of new technology should not be steered solely by companies like Google or venture capitalist-funded R&D. Autonomous vehicles may radically alter our transportation landscape. But their impact on car culture, as we know it, should be part of a public policy discussion that addresses the issue of legacy vehicles. That conversation must address what types of vehicles will be allowed where. If we want to keep the street open for rodders and highway cruisers, then we must insist on a place at the table.

I can't help but think about what some horse buggy craftsman must have been thinking when he saw the first of Henry Ford's Model T's roll into town in 1908. They likely reacted with some wonderment, but I doubt there was a "this changes everything" revelation. Yet by the end of the Model T's production in 1927, more than 15 million had been sold. And everything *had* changed. We are, I'm convinced, at a similar pivotal moment.

Or maybe not. For the life of me, I cannot see rodders giving up their steering wheels. Hot rodders drive what they build; that's the point. Granted, some buy their rides, but everyone that I met during my odyssey loves to drive them. On the other hand, transportation planners and venture capitalists see the self-driving car as an alternative to the hassle, congestion, and danger of driving.[5] So they intend to take the human element (as in "driver") out of the equation. It's two very different cultures.[6]

My own perspective is that blind faith in technology can lead to terrible consequences. Consider the fate of the unsinkable *Titanic*, four years after the introduction of the Model T. Not to mention the host of unimaginable events, right up to the 2011 Fukushima nuclear disaster or the 2015 Southern California Porter Ranch gas leak. Evidence abounds that our faith in technology has, with some regularity, been tragically misplaced.

Of course, some argue that these events are accidental one-offs. And that controls are now in place, or will shortly be put in place, so things like this will never happen again.

Yep. Humor me for a moment. Imagine that driverless cars soon fill our roadways. That's the vision of venture capitalists, positioning to make a financial killing on the transition. Most of their visions require a Skynet-managed robocar-based transportation system. What could possibly go wrong?

Lots. Every hot rodder will tell you that if something can go wrong, it will. This is the generalized version Murphy's Law. Personally, I prefer the more elegant "the perversity of the Universe tends towards a maximum." Take your pick; here are a couple scenarios. Imagine Southern California's Big One; the earthquake splits a section of an overpass. And driverless robocars plunge, like lemmings, to the ruined roadway below. Screaming passengers are unable to bail out before going over the edge because the doors are always locked when the vehicle is in motion. Or imagine the sudden appearance of a sinkhole in the middle of a Florida highway. At a grander scale of universal perversity, a large solar flare accompanied by coronal mass ejection (CME) would severely damage GPS signals, wire and wireless transmissions, and the power grid.[7] Or circuit boards and chips could be fried by a nuclear detonation-generated electromagnetic pulse (EMP).[8] The next wave of technology may eliminate the hand-driven vehicle but set up our transportation grid for catastrophe. Nature always sides with the hidden flaw.

It makes sense that many elements of contemporary car culture need to remain on our roads and highways. When things go south, hot rodders will be able to build, fix, and run rides that work. Survival will not be enhanced by mechanical cluelessness.[9]

24

Final Thoughts

You don't need a book to help you remember the hot rodding you knew and loved. Our rides were sweethearts of yours and mine, hot cars with many lovers. Car culture, as we knew it, is passing away but we still love her. The transportation grid re-programmers will bury her without flowers, but our rides will live on in the hearts of their lovers.— My reworked version of book inscription (with apologies) by Charlie Russell[1]

When I started this book in 2008, gas prices were above $4 a gallon and seemingly poised to run higher. Combined with the graying of our cohort and the negative environmental impacts of hot rods, I imagined that we were nearing the end of car culture as we know it. I had also just finished Charlie Russell's *Trails Plowed Under*, a book of warm and nostalgic stories about the Old West. Running throughout the book is a melancholy feeling for the passing of that era. In hindsight, I guess that I was transferring some of that sadness to my thinking about the future of hot rodding. I'd even considered titling this book *The Doom of Zoom*.[2]

But who knew that we'd generate a fracking-based oil glut while I was working on this book? Hot rodding seems to have gained a new lease on life. Still ahead are environmental issues society must address, the rise of autonomous vehicles, and the aging out of baby boomers. But many conversations with gearheads have convinced me that rodding, in some recognizable iteration, will persist. The question is for how long.

I concede that I am not exactly batting a thousand with my predictions. Nevertheless, based on U.S. Census data, it is clear that hot rodding, at least as we know it, is approaching a tipping point. Consider that the number of baby boomers has been decreasing since 2012. And the pace of this decline is accelerating. There were 77 million baby boomers when the first of us turned 65. By 2030, boomers will be between 66 and 84 years old, and their numbers will drop to 60 million. Half will be 75 or older. In 2060, baby boomers will number just 2.4 million, every one of them 96 or older.[3]

Mortality is a bitch. But there's no fighting it, and that means hot rodders will be moving into old ages. Based on my interviews, many boomers "retire" from hot rodding sometime in their mid-70s. There will be some diehards, of course, but as boomers age, their mobility decreases. And friends pass or move away. Aging rodders attend fewer and fewer events. Rides sit idle in the garage. And because (as one rodder recently reminded me) retirement is expensive, many cars will be sold for living expenses or to pay for health care.

24. Final Thoughts 179

Gene Winfield shop display in the Petersen Automotive Museum (March 2013, Los Angeles, California).

So the activity as we know it has an expiration date. Nevertheless, despite the decline in numbers, the cohort will continue to play an important role in the activity. This will extend perhaps through 2029, when all of the boomers will be over 65. Beyond that date, hot rodding will be morphing into another form that only modestly reflects today's car culture landscape. As I stated in the first chapter, the hot rodding landscape is tightly linked to the baby boomer cohort. These guys were kids when their fathers and uncles joined the culture after their return from World War II. Today's hot rodders became gearheads when an older family member or friend took an interest in them, taught them wrenching, triggering the expression of their gearhead gene.

Hot rod culture really began to spread after men returned from the war. They'd seen enough; they just wanted to settle in and have a good life. But for many of them the good life wasn't good enough. Wrenching on an old car provided a socially acceptable, solitary opportunity to focus on something tangible. It was a form of therapy for many. And the outlaw street racer nature of rodding offered many of them the shot of adrenalin that they craved as a tonic for the mind-numbing normalcy of everyday life. The unregimented element of the activity was a great attraction. For some, it was the only way to keep moving forward. As time passed, they found similar souls and formed car clubs. As their kids or neighborhood kids showed some interest, they took them under their wing.

Based on my research, this is a good partial thesis for the rapid post–World War II evolution of hot rodding. Of course, from the earliest days there was car racing, and the need for speed.[4] Out West it was dry lakes racing; in the Midwest and back East it was dirt track. And street racing—sanctioned and not—was universal. But it was not all about speed. Style has always been important. Hot rodders have always appreciated good proportions, efficient design, and style. From the earliest days, this was how rodders rolled.

Baby boomers, then, are really second generation hot rodders. And there seem to be spatial and temporal geographies to the hobby.[5] An analysis of car show registration data would show correlations between cars of certain eras (traditional hot rods, fat fendered street rods, muscle cars, and so on) to certain age groups. I would expect to see that boomers favored the hot cars at the time they were in their teens. Or, perhaps they preferred the style of their first or second vehicle. Back in the day, hot rodders drove what they could afford: mostly older cars, modified and customized to differentiate them from other older daily drivers. Some may have rodded their cars as part of a pissing contest to establish their place in the pack. Others likely cobbled together whatever they could just to have a ride.

You drove what you built. Most rodded their cars for cruising and, perhaps, street racing. And there was a courting element to the activity; a good flame job made your car (sorry) hot. Hot cars attracted mates. The Friday and Saturday night endless laps around drive-in restaurants offered an opportunity to strut your stuff. There was a method to the madness. Many of today's boomer rodders look back with nostalgia at those *American Graffiti* days. Key elements of identity were formed and a love for cars was permanently tattooed on their souls.

25

At the End, 2017

> *Time stalks all car guys. Rust works its implacable corruption; rubber and plastic crack and split; fluids go dry. Keeping these machines on the road demands a set of increasingly antiquated skills and an abiding tolerance for breakdowns. Even the most badass of Camaros can't outrun fate, and neither can their drivers. Collector magazines and sites are full of estate-sale listings for unfinished projects, each a bittersweet score for a new owner.* —"American Beauties" (*AARP Magazine* story by David Dudley about Boomers buying their dream cars at Barrett-Jackson auction)[1]

Dudley's statement pretty much nails our circumstances. It's not the final nail in the coffin of hot rodding, because there are younger guys out there taking on project cars, restoring and maintaining rods and customs, and participating in the culture. Here's the issue: Younger guys are not joining in sufficient numbers to replace those aging out. We find ourselves in this circumstance because of the demographic shift that is beginning to hammer on hot rod culture.[2] Baby boomers, the mainstay of today's hot rod culture, are moving into old age. And the last time I looked, retirement communities and nursing homes don't have man caves, TIG welders, engine hoists, or paint booths. Like it or not, old rodders lose traction. Then they die.

What does this look like at ground zero? Upstate New York, where I live, had the Cortland Antique Automobile Club. For many years, the group held a car show at a local park. Over the past 20 years, it morphed into a hot rod, street rod, and muscle car show. During that time few, if any, antique cars showed up. The club's brass era (pre–1916) vehicle trophy was rarely awarded. The club went belly up as most of the membership aged out of the activity. We've lost the people with the skill sets and the interest necessary to maintain these vehicles. It catches up with us; rust never sleeps.

Then there's the question of whether old men and old metal have any place in tomorrow's transportation landscape. But I won't beat any more on that horse. From a demographic standpoint, my best guess is that we've got maybe 20 years before it all goes to hell. At that point a majority of baby boomers will have cashed out of the activity. Nostalgia-based vehicle values will crash. Of course, this represents an opportunity for younger guys to pick up some great old metal. But how many of the Gen X, Gen Y, or Millennial cohorts have any interest in the activity? How many young kids—say from the age of eight to fourteen—will be taken under the wings of gearheads?

So who are tomorrow's gearheads? How will antiquated skills persist? What does hot rodding look like 30, 40, or 50 years out? What's left behind and what's carried forward? My darkest vision is that there may only be a group of traditional rod and custom guys who keep things going in a manner like the Civil War reenactors. But even these reenactors may one day be unable to muster enough participants to reenact Pickett's Charge at Gettysburg. Old age inexorably takes its toll. Things roll on, regardless of our current wants and desires.

It's also hard for me to imagine that hot rodders will be able to continually stick a finger in the eye of progress. Maybe they can give it the finger for a while; but like most things, hot rodding has a temporal component. As time rolls on, there is less and less *stopping* to master antiquated skills. Less is carried forward. We can put on a brave face, but in the end if we do not recruit new enthusiasts, hot rodding will not survive.

The future typically blindsides us. It's hard to say how environmental concerns will factor into all this. It's not only what we don't know that gets us; often it's what we think we know that turns out to be wrong. Remember that "the perversity of the Universe tends towards a maximum." I may be wrong about how hot rodding rolls on. There are vibrant online forums like the H.A.M.B. or BangShift[3] that serve as a repository of history and knowledge about traditional techniques, standards, and design parameters. And there are a handful of builders like the guys at the Rolling Bones Hot Rod Shop in Greenfield, New York, who exclusively build cars that "capture the look, the feel, and the soul of a late forties–early fifties hot rod." Perhaps these brain trusts will keep things rolling.[4]

Abandoned homestead and vehicle—possibly a '35 Dodge Touring Sedan—at Marysville, Montana (July 2008, photograph by Crystal Chase).

Future technologies may revolutionize our definition of the enterprise. Down the road, individuals may customize their rides with an app. Perhaps digital "paint" will allow for instant color changes with a few pushes on a touch screen. And rodders can select digital flame jobs or classic pinstriping from an extensive menu. Maybe new panels for virtually any vehicle will be created using large format 3D printers. Perhaps chop, nose, and section jobs will be done on the computer and body panels printed out, paint ready; instead of building, it is assembly. Imagine how easy it would be to alter on-screen a wheel opening or move a headlamp mount. All with just the touch of a finger.

Maybe rust *can* be put to sleep, thanks to new age composite body panels and chassis components. Things may get crazy from a design and build standpoint. The old stuff could be recreated—an endless supply of the real classics—and stretched as a skin over new electric running gear. You'd get infinite choices, speed of assembly, and flat-out speed. Imagine the scene when traditionally inspired, period-accurate, next-gen speedsters hit the Bonneville Salt. Or when body panels and components can be printed out for a high tech electric version of Hirohata's '51 Merc that captures all the rolling poetry of the original.[5] How does any of this relate to hot rodding as we know it? Maybe it's a stretch; perhaps whole different skill sets will be required. But maybe this has to happen if the tradition is to be carried forward.

At the more "crazy shit" edge of speculation, perhaps the whole enterprise morphs into a virtual hobby. As increasing pressure is brought to bear on rodders, as their unscripted activities are effectively constrained by regulations, and as the middle class withers away, there may be no place in the real world for hot rodders and old metal. Maybe future rodders will experience what passes for the open road only by going virtual. Hot rodding? There's an app for that. Perhaps there will be an app for assembling a ride, showing it at a virtual show and racing it on a virtual street or dry lake. Maybe, God forbid, all that rodders are left with is a new-age style of bench racing.

Of course, this is just speculation. But I wonder: Does hot rodding morph into something that today's boomers would find hard to recognize? Do old men and their old metal simply rust and fade away? Are we approaching the end of days for hot rodding? The only thing that is certain is that the hobby will surely wither away unless more of us take young kids under our wings.

I began this book in 2008. The first national show that I attended was the 4th Annual Goodguys Great American Nationals at Pocono Raceway. Gary Meadors (R.I.P.), founder of Goodguys, set me up to shadow Bill Goodwill, who was in charge of the awards team. That weekend, the sky opened up and the rain came down so hard it was like a cow pissing on a flat rock. So Bill suggested that I shadow him at the upcoming Nashville show, which I did. Since then, I have put nearly 40,000 miles on my '58 Apache, going to shows of all sizes. I have completed two Power Tours, a Street Rodder Tour, and driven twice from Upstate New York to the West Coast and back. I have covered shows and written articles for *Custom Classic Trucks* magazine.

During this odyssey, I have met and interviewed many rodders. I enjoyed great conversations, and I am profoundly indebted to everyone who shared their thoughts about the future of our enterprise. It would be an understatement to say that my thinking about "how we roll" has taken many twists and turns. Thoughtful members of our community have stretched my notions about the future of hot rodding many times.

So what does hot rodding look like 20 years from now? Tom Vogele, editor of *StreetScene Magazine*, read some draft chapters of this book and commented that

> In general, I agree with your thoughts but I strongly believe some form of personal transportation will always propel people to modify that form of mobility. The main point is that the age of those who choose to modify their car changes over the years. They will modify cars that they relate to from an earlier age. Many 20–30 year olds I talk to think Mustangs and Camaros are what they view as "old" cars. But I hold out that the hot rods as we have known them will hold their appeal for many years into the future, albeit maybe not in the same numbers we enjoy today. There is no doubt there is a visual and real movement upward in the years of cars being modified.[6]

I agree that traditional hot rods will maintain their appeal for many years. And that there will always be some compelled to modify their vehicles. So it seems safe to say that elements will be carried forward. Nevertheless, I am convinced that new technologies will ultimately push aside many of the traditional ways and means of the craftsmanship that today's hot rodders incorporate into their rides.

New school will trump old. Change happens. Obviously, there will be holdout by some traditionalists: God bless them. But the tools and tooling will change in ways that we cannot anticipate. And evolving transportation policies related to driverless vehicles will impact just how much of the road remains open to car guys. If opportunities do not remain for expression of the gearhead gene, then the old metal will be scrapped or dragged back into the weeds. And the passion for our hobby will be slowly strangled by interests that compete more effectively for the next generation's attention.

At the moment, we are a "boomer-fueled" activity. Go to any show and you will see mostly graybeards. And good numbers of them on what are euphemistically called "mobility scooters." So while our passion for hot rodding remains undiminished, our collective numbers will decline. We are also running against the currents of technology, environmental regulations, and the increasing regimentation of everyday life. The open road is closing. We are approaching the end of an era.

So is the message of this book one that today's hot rodders can roll with? I had several of my gearhead friends review draft chapters of this book as a check on my thinking because I was concerned that my analysis was one that many hot rodders might not be willing to hear. One of my reviewers was Gary Buehler, whom I first met at Goodguys Pocono in 2008. I happened to sit beside him on a golf cart under the race stands while it rained heavily. We started talking, hit it off, and have been good friends ever since. Gary is a graybeard boomer, a master fabricator, and a hard core, traditional-style hot rodder. After reading my manuscript, Gary sent me the following message

> Heady stuff here ... all of it! On the scale of 1 to 10, it is a 13! I'm thinking and it hurts. Every car guy I know should read this.... I'm 75 looking down the barrel of 76, so your point about silverbacks crapping out mid 70's and for sure by 80, scares the hell out of me. In the last three months, I've said to my wife at least three times, "Maybe it's time to sell the rods and, you know, maybe get a new Vette. I really like the looks and style of the new ones." To which she replies, "Really? You don't think you'd miss the hot rod?"
>
> Last fall, I drove the '30 Model A tub over to my son's house for storage for the winter. Mother's Day, he was here and said, "Dad, I replaced the rear wheel seal, checked the car out and was going to take it for a ride but noticed it wasn't currently registered." I gave him the new registration and insurance card. He loves that car. I know it won't be back in my garage

25. At the End, 2017

… too noisy, too loud, I don't want to get wet any more, too hard to steer, too hard to stop, it's very kool and it's very uncomfortable.

But Gary goes on to say that he is building an "Old School" Model A coupe with his now 15-year-old grandson. It is a three-year project to be finished next year. About that effort Gary writes,

> What the hell am I thinking? It's a 4-speed. My grandson doesn't even have his permit. He will learn to drive in a modern car. The hot rod doesn't have anything a modern car has … power steering, power brakes, and power anything … and drums brakes! No air bags! Eighty-year-old engineering! The car isn't safe! He's my grandson.

So what the Hell is Gary thinking? Why is he building a rod with his grandson? He is passing on his passion for hot rodding and his knowledge of how everything on a hot rod works. His grandson can now

> …weld, paint, sand, fill, sand, paint, wire, design and lay out a dash, mount tires, build runners for a seat, put glass in, wire tail lights and install them where there were no tail lights before. He can put in roof ribs, install a windshield, pack wheel bearings, install backing plates, wheel cylinders, attach brake shoes, and adjust them. He can flare brake lines, bend them, attach them to the master cylinder, and add a proportioning valve. He knows what

Gary Buehler and grandson Gavin working on their "Old School" Model A coupe on Labor Day 2016 (Pultneyville, New York. Photograph provided by Gary Buehler).

.040 over means. He picked out the gauges and wired them. He picked out the valve covers. He understands ... what each component is for and how it all works. And maybe, just maybe, he will remember his grandpa.[7]

Gary's reaction to my manuscript addresses my central point. The scale of our enterprise will decline as boomer numbers decrease. But the gearhead gene will not be totally clipped from our DNA as long as rodders continue to pass on their passion for building something with their own hands. Our non-conformity can persist; there is a way to roll forward.

When I started this book in 2008, I knew that I had a lot to learn about hot rodding. Even now, after more years than I ever imagined the effort would take, I still have more to learn. That being said, the book that you are holding is the book that I was compelled to write, based on conversations with hot rodders all over the country. While some rodders may remain in denial, most are pretty sharp people willing to keep an open mind; they will give these ideas a fair hearing. That's all I can ask.

There will be a good deal of geographic variation in the timing of how this all plays out, but I do think that I have much of this right. We are beginning to see discussions about the graying nature of our enterprise on car and hot rod related message boards. For example, in response to a recent (2/2017) H.A.M.B. thread titled, "So ... at what age did you have to give up building hot rods?,"[8] Randy Haugen (v8deuce) observes that *"Seventy-five is a good average age for what a guy can expect to be able to work to and actually get something done."* He then says to get a tape measure and pull it out to 75 inches. Then, assuming each inch is a year, put your finger on your age. Randy writes that

> This will put in perspective how much time is left. When I did this, my priorities came into focus immediately; now I'm getting my bucket list done and enjoying my time with my grandmonsters who mean the world to me. Who knows for sure when the grim reaper will show up but for now I'm done with the stress of paying for stuff that only impresses the neighbors. Now is for me and the ones I can count on. As for working on cars, I hope I go working on one with my family around helping.

Finally, my friend Steve Chase reviewed draft chapters of this book. In a note about what I sent to him he wrote,

> Most of the old guys I know are lamenting what you describe. We are losing them to the lofty cruise-in above (or below) depending on who is talking. However, it reminds me of the racecar drivers my dad ran with, in that if one died they knew it could be them but they kept going and justified it as a tribute to those that went before.[9]

I can ride with that. However this sorts out, hot rodding will continue to move forward. But as I said at the beginning of this book, rust never sleeps. Almost everything, from a personal transportation standpoint, is going to change. And gearheads are racing against a very stiff head wind generated by changing demographics, technological innovations, and environmental imperatives. Other interests seem destined to compete more effectively for the attention of future generations. We are approaching the end of an era.

Whatever the road ahead, based on my eight year odyssey of road trips and hundreds of conversations, I conclude that hot rodding is today a vibrant American subculture. And this book celebrates the persistence of rodding culture as a revolt against the regimentation of everyday life. Gearheads abide. They design, build, run, fix, and rebuild their rides. As long as there are open roads, hot rodders will defy the trend toward mechanical cluelessness, anonymous styling, driverless vehicles, and a "get back in line and wait to be served" mentality that permeates our current transportation landscape. That is how we roll and why it matters.

Chapter Notes

Chapter 1

1. Tom Vogele, *StreetScene* [NSRA monthly magazine]. December 2015, p. 8.
2. The terms driverless and autonomous are used interchangeably by many authors. The lawyers will sort this out. One legal firm's website says that an "autonomous … car is one that is responsible on some level for driving the car automatically. This description can apply to old technologies like cruise control, newer technologies like automated parking and responsive lane awareness, and developing technologies like fully automated driverless cars. Driverless cars, on the other hand, are defined much more simply. A driverless car does not require a human driver. Everyone inside a driverless car is a passenger." See the Bailey and Oliver law firm blog of February 26, 2018. "Autonomous vs. Driverless Vehicles." At http://www.baileyoliverlawfirm.com/news/2016/feb/26/autonomous-vs-driverless-vehicles/.
3. Andy Sheehan, SS Garage. "Ford Focus RS Proves Cars are Still Alive." February 10, 2015, at http://www.streetsideauto.com/blog/car-enthusiasts/ford-focus-rs-proves-cars-still-alive/.
4. The dice were cast, in my case, and things came out snake eyes. I lost my '58 to a garage fire in August of 2016. Probably due to mice chewing on the under dash wiring. Fortunately, the fire was confined to the cab and surrounding area but the truck was a total loss. Obviously this was, and remains, a heartbreaking circumstance.
5. Maybe not so far down the road. See Alex Nishimoto, "Kids Born Today May Never Drive a Car." January 3, 2017. http://www.automobilemag.com/news/robotics-expert-predicts-kids-born-today-will-never-drive-car/.

Chapter 2

1. See https://en.wikipedia.org/wiki/Rust_Never_Sleeps.
2. See http://collection.mam.org/details.php?id=19229.
3. "Eastman Johnson's Paintings Shown." *New York Times*, February 11, 1907, p. 9. For more information on Johnson, see http://www.nytimes.com/1999/10/29/arts/art-review-capturing-the-moods-of-a-nation.html?pagewanted=all.
4. See https://www.tuckersparts.com/Cab-Assembly-CAB-5559.html.
5. Personal communication, February 12, 2008.
6. Titus Livy (59 BC–AD 17), Roman Historian. Livy also observed that "Men are seldom blessed with good fortune and good sense at the same time."

Chapter 3

1. John Jerome, *Truck*, p. 176.
2. David Miller, *The Complete Paddler: A Guidebook for Paddling the Missouri River from the Headwaters to St. Louis* (2005).
3. This swap is also featured at the Street and Performance website at http://www.hotrodlane.cc/77chevytruck/77chevy.htm.
4. Pers. Comm., 12/08/07.

5. My friend and editorial consultant Roger Jetter calls this the "Old Hot Rodder's Nemesis." As he states it, "When changing or modifying things on any vehicle, one must change several other things to accommodate the one change" (pers. comm., 1/17/17).
6. The Cayuga Performance Auto Machine Shop is located in Newfield, NY.

Chapter 4

1. John Jerome, Truck, p. 174.
2. See De Graaf, J. and Wann, D. and Naylor, T.H. *Affluenza: The All-Consuming Epidemic.*
3. See http://www.nytimes.com/2008/04/22/science/22conv.html?_r=0; Also see Daniel Todd Gilbert, *Stumbling on Happiness.* I also recommend Gilbert's TED talk at http://www.ted.com/talks/dan_gilbert_asks_why_are_we_ happy?language=en.

Chapter 5

1. Studs Terkel, *American Dreams, Lost and Found* (1980). The "no hyperbole" quote is from Terkel's *Acknowledgments and Apologies* section, p. xvi. The "voices captured by hunch, circumstance and rough idea" quote is from the *Introduction*, p. xxv.

Chapter 6

1. The All America Racing Barracuda was produced only in 1970. Only about 1,500 were built. See http://www.stockmopar.com/aar-cuda.html.
2. The International Show Car Association, see http://www.theisca.com/?page_id=34.
3. For context about the AMBR award see the Brian Brennan's *Hot Rod Network* article, "2016 America's Most Beautiful Roadster is … Darryl Hollenbeck's Traditional 1932 Ford Highboy" at http://www.hot rod.com/events/1601-2016-americas-most-beautiful-roadster-is-darryl-hollenbecks-traditional-1932-ford-highboy/.
4. Rat's Glass is one of the pioneer makers of fiberglass street rod bodies. See http://www.ratsglassbodies.com/.
5. Darryl Starbird's National Rod & Custom Car Hall of Fame Museum is in Afton, Oklahoma. See http://www.darrylstarbird.com/museum.htm.

Chapter 7

1. See Rik Hoving's tribute to Bo, "Rest in Peace Bo Huff," at http://www.customarchronicle.com/in-memoriam/rip-bo-huff/.
2. For more information about Bo Huff's '57 Ranchero, see discussion in "The Hokey Ass Message Board" started by Jonnie King, Nov. 10, 2011, at http://www.jalopyjournal.com/forum/threads/bo-huff-building-the-57-panochero.642724/.
3. See a photo of Bo's '42 Ford "the Rockabilly Hound Dog," at Ron Springer's Kustom Car Parts website at http://kustomcarparts.com/drive-that-kustom/.
4. See the interview at https://www.youtube.com/watch?v=5M6EhgEJma4.
5. See the John D'Agostino's Celebrity Kustoms website at http://celebritykustoms.com/.
6. See the Viva Las Vegas Rockabilly Weekend Show website at http://www.vivalasvegas.net/car-show.
7. See http://www.chopitkustom.com/Home.html. I met Gary and his family while visiting with Bo Huff. Sadly, Gary—a truly gifted Kustom designer—passed away May 30, 2016. Also see http://www.rodauthority.com/news/kustom-builder-gary-chopit-fioto-dies-of-heart-failure/.

Chapter 10

1. See Dakota Wentz's article, "Stardust: A '50s Custom Meets the 21st Century" in *Custom Classic Trucks*, July 2012, pp: 36–40.
2. See Starbird and Bledsoe's book, *Darryl Starbird: The Bubble Top King Custom Car Creations, 2006.*

Chapter 15

1. The Galaxie 500 XL two-door Club Victoria sold for $3,222. The convertible sold for $3,787, which would be about $30,000 in 2017 (calculated using the Dollar Times Inflation Calculator at http://www.dollartimes.com/inflation/inflation.php?amount=1000&year=1964.) According to the NADA Guide, today's average selling price for the two-door model is $15,900.
2. Fred C. Offenhauser, founded Offenhauser, an early manufacturer of high quality automotive performance products. See http://offenhauser.co/.
3. Roger Jetter is the author of a series of books focusing on what some might say was his misspent youth as a hot rodder. Those of us who know him, know better. He's still a rodder, a skilled metal fabricator, and a first-rate story teller. For many years he was a monthly columnist for *Goodguys Gazette*. See his website at: http://rajetter.com/books/.
4. Rod is referring to an incident in the 1955 movie *Rebel Without a Cause*, which has a racer driving off a cliff when he is unable to bail out from his car because his jacket sleeve gets caught on his door latch handle. There are many myths associated with Dean's September 30, 1955, death. He was 24 when died in a car accident at an intersection in Cholame, California.

Chapter 16

1. A NACA inlet is a low-drag air inlet design, originally developed by the U.S. National Advisory Committee for Aeronautics 1945.
2. A back-halved chassis has its floor cut out from the firewall back and a completely new suspension installed to create a full-on drag car. To visualize this see, "1984 Camaro Drag car build," at http://www.thirdgen.org/forums/fabrication/606908-1984-camaro-drag-car.html.

Chapter 21

1. Bob Lutz, "My Turn: The Road Ahead for Cars." *Newsweek*, May 5, 2008.
2. CNN Money Special Report, "Gas price record reaches $4 a gallon." June 8, 2008. http://money.cnn.com/2008/06/08/news/economy/gas_prices/.
3. Nick Bunkley, "Detroit Automakers Compete for a Vanishing Truck Market," *New York Times*, June 5, 2008. http://www.nytimes.com/2008/06/05/business/05auto.html. Also see Bunkley, "Highly Rated Auto Plants Set to Close," June 6, 2008. http://www.nytimes.com/2008/06/06/business/06auto.html.
4. I verified the 27 mpg at the Watkins Glen Eco-Challenge, which had participants weigh their vehicles before and after traversing a measured course. Speed was limited to 60 mph. I did not dog the event and still got this respectable mileage.
5. "Who took the wheels off your American Dream?" Chrysler advertisement, *Barron's*, May 19, 2008, p. 52.
6. See Brian Brennan's article, "The Defining Moment: Hot rodders and the law" in *Street Rodder*, April 2010, pp. 118–122. Although he focuses on the issue of fraudulent registrations, Brennan discusses hot rods and emission standards. He observes, "…the day may not be too far out there when emission standards will be mandated."
7. Upon reading my concerns about rodding being strangled at $5 or $6 a gallon gas prices, my friend Roger Jetter observed that, "I think $4.00 a gallon … was the tipping point … not [Lutz's] $10.00 a gallon. I know people (rodders) that simply won't play at that price! Hell, they won't even pay $15.00 to go to a local across the city rod run!" (Pers. Comm., January 17, 2017).
8. For more information on externalization of costs, see search results for the term at https://wiki.p2pfoundation.net/.

Chapter 22

1. See http://www.ideoautomobility.com/.
2. For additional context, see Colby Martin's (SEMA) Legislative Scene editorial, "Filling the Generation Gap Solving the 'Classic' Car Question: Who and What's Next," in *StreetScene*, Volume 45 Number 8; August 2015, pp. 138 and 140. Martin observes, "Over the years, I've noticed a lingering anxiety among members of the car community regarding the makeup of the next generation of enthusiasts and what they will drive/collect."
3. See Parsons, J., Padilla, C., and Arellano, J.E. (1999), *Low'n Slow: Lowriding in New Mexico*. Santa Fe, NM: Museum of New Mexico Press.

4. Yonah Freemark, "Will autonomous cars change the role and value of public transportation?" at thetransportpolitic.com/.http://www.thetransportpolitic.com/2015/06/23/will-autonomous-cars-change-the-role-and-value-of-public-transportation/.

5. Dan Weikel, "Driverless vehicles and the future of L.A. Transportation." Los Angeles Times, November 29, 2015. http://www.latimes.com/local/california/la-me-california-commute-20151130-story.html.

6. Matthew Phenix, "Ten things the driverless generation will never experience: Autonomous cars will do for traditional cars what traditional cars did for the horse and buggy." Bbc.com, January 22, 2016. http://www.bbc.com/autos/story/20160122-lost-in-automation.

7. Sarah Hunter, "Transport Innovation Talks." May 28, 2015. [Sarah Hunter (Google [x]): Transport Innovation Talkhttps://www.youtube.com/watch?v=UwxeyNlKT1A. Every hot rodder should watch this video.

8. Arthur St. Antoine, "2016 Automobile All-Stars: The Winners ("From a field of this year's most compelling new cars, we pick the seven that most closely match our ethos"). *Automobile*, March 22, 2016. http://www.automobilemag.com/news/2016-automobile-magazine-all-stars/.

9. Martin cites an MTV report finding that Gen Y and Millennial drivers "...continue to see car ownership as a way to establish independence and shape their unique adult identity. In fact, 75 percent would rather give up social media for a day than their cars, and 72 percent said they would rather give up texting for a week than their cars" (Martin, *op. cit.*, p. 138). If this information is accurate, perhaps things are looking up from a "car guy" perspective. Of course, things would be also looking up if younger drivers would give up texting while driving.

10. I shared a late draft of this chapter with Chris Gerdes, a professor of mechanical engineering and director of the Revs Program at Stanford University. He responded with the observation that "I think you touch on something quite deep with the difference between driving and riding. One of the things I have discussed with one of our program sponsors is whether going from moving ourselves in the world to being moved in the world changes our sense of agency in a fundamental way. If we no longer move ourselves in the world, do we start to think differently about how we can shape the world? Does this start to translate to more apathy or acceptance of institutions? Important questions to consider, I think" (pers. com. 3/16/2017).

Chapter 23

1. This was Segment 10 of show #146 (The Model A Prius) that originally aired on 11/12/2011. You can listen to Tom (RIP) and Ray snort and guffaw through this episode at http://www.cartalk.com/content/1146-model-prius. Many NPR stations continue to feature past episodes in a program called *The Best of Car Talk* at http://www.npr.org/podcasts/510208/car-talk. Gearheads will especially appreciate program #1648: The Black Hole of Auto Metaphysics (November 26, 2016) where the boys discuss a postulate that a non-essential repair on an old car creates a parallel universe where everything else starts to break.

2. Jason Udy, "Tesla Model X P90B is the Quickest SUV We've Ever Tested." *Motor Trend*, March 23, 2016. http://www.motortrend.com/news/2016-model-x-p90d-numbers.

3. Camaro SS coupe performance data from GM website. http://media.chevrolet.com/media/us/en/chevrolet/home.detail.html/content/Pages/news/us/en/2015/sep/0914-camaro.html]. 2016 performance data at http://media.gm.com/media/us/en/chevrolet/home.detail.html/content/Pages/news/us/en/2015/sep/0914-camaro.html.

4. It won't be long before some stand-up comedian riffs on the future perils of the robocar. Will friends let friends ride drunk? Or, how a night of hometown Columbus, Ohio, adventure might end with the pressing the wrong button and waking up in Homer, New York.

5. It will be a while, if ever, before driverless cars eliminate danger. Although Tesla CEO Elon Musk is bullish on the technology ("I believe that it's probably better than human at this point, in highway driving. Or certainly will be as street learning gets more and more sophisticated"), things are far from foolproof. See Edward Loh, "Latest Tesla Update Lets You Summon Your Car from Wherever You Are." *Motor Trend*, January 10, 2016. http://www.motortrend.com/news/latest-tesla-update-lets-you-summon-your-car-from-wherever-you-are/. Also see Bob Sorokanich, "Police Say Laptop, DVD Player Weren't Running in Fatal Tesla Autopilot Wreckage." *Road & Track*, July 7, 2016. http://www.roadandtrack.com/new-cars/car-technology/news/a29877/tesla-model-s-autopilot-fatal-crash-laptop-dvd/ and Bill Vlasic and Neal E. Boudette, "Self-Driving Tesla Was Involved in Fatal Crash, U.S. Says. *New York Times*, June 30, 2016. Tesla's response, "This is the first known fatality in just over 130 million miles where Autopilot was activated,") is at Tesla's website, https://www.teslamotors.com/blog/tragic-loss.

6. How different are the "self-driving" vs. "driver" visions? Consider "2017 Fiat 124 Spider Abarth vs. 2016 Mazda MX-5 Miata Club," by Todd Lassa, *Motor Trend Newsletter*, July 18, 2016. These cars, he writes, "are the best antidotes for the coming onslaught of autonomous vehicles. No lane departure controls, no cross-traffic alerts, nor blind-spot information systems. Say 'hell no' to automatic braking systems. Autonomy? We don't need no stinkin' autonomy. [These] are cars to drive, not to be driven in." At http://www.automobilemag.com/news/2017-fiat-124-spider-abarth-2016-mazda-mx5-miata-club/.

7. For fun reading, see Sebastian Anthony's "The solar storm of 2012 that almost sent us back to a post-apocalyptic Stone Age." Extremetech.com, July 24, 2014. Make sure to look over the comments for additional context. http://www.extremetech.com/extreme/186805-the-solar-storm-of-2012-that-almost-sent-us-back-to-a-post-apocalyptic-stone-age#disqus_thread.

8. R. James Woolsey and Peter Vincent Pry, "The Growing Threat from an EMP Attack" *Wall Street Journal* August 12, 2014. http://www.wsj.com/articles/james-woolsey-and-peter-vincent-pry-the-growing-threat-from-an-emp-attack-1407885281. Again, read the comments for additional perspectives.

9. With respect to the notion of mechanical cluelessness, I recommend that you read Matthew Crawford's book, *Shop Class as Soulcraft: An Inquiry into the Value of Work*. Crawford argues that we have lost our fundamental manual competence, that we cannot fix our stuff, and increasingly steer our kids away from occupations that get their hands dirty.

Chapter 24

1. Russell's inscription is mentioned in the footnotes of B. Dipple's "Introduction to the Bison Books Edition" in *Trails Plowed Under: Stories of the Old West*, p. xix. Bison Books, 1996. Dipple credits Frederic G. Renner's *Paper Talk: Illustrated Letters of Charlie Russell* (Amon Carter Museum; 1962) for the quote.

2. Doom of Zoom, © David L. Miller. I did not use this title for a variety of reasons, principally because hot rodding is not just about speed. Still, it's hard for me to imagine that whatever form "personal conveyance devices" take on, there won't be some very clever minds focused on tricking them out to go faster.

3. These data are found in the U.S. Census Report, "The Baby Boom Cohort in the United States: 2012 to 2060," by Sandra L. Colby and Jennifer M. Ortman, May 2014. Pp. 25–114. See Current Population Reports Population Estimates and Projections section, https://www.census.gov/prod/2014pubs/p25-1141.pdf. For additional context, see, "Fueled by Aging Baby Boomers, Nation's Older Population to Nearly Double in the Next 20 Years," Census Bureau Report," May 6, 2014. http://census.gov/newsroom/press-releases/2014/cb14-84.html.http://www.census.gov/newsroom/press-releases/2014/cb14-84.html.

4. One of my favorite books is Pat Ganahl's *Hot Rod Gallery*, which traces the history of hot rodding from the pre–World War II beginning to 1960. The photos and commentary in this book are outstanding. Pat's books, along with those of Bo Bertilsson, Gerry Burger, Tom Cotter, Dain Gingerelli, Scotty Gosson, Ken Gross, Bob Larivee, Sr., Alan Mayes, Andy Southard, and Peter Vincent, are outstanding sources of information about hot rodding. The works of many of these authors show up in *The Rodder's Journal* and are listed in catalogs of curated books they offer to enthusiasts. TRJ represents the gold standard for writing and photography focused on hot rodding, which they state is "one of the world's most unique and passionate pursuits." Obviously, I concur.

5. I began thinking about this after reading Brian Brennan's *Street Rodder* editorial, "Is The Early Hot Rod Dead?" March 2014, p. 10. Although not discussing the spatial and temporal geographies of the hobby, Brennan notes the rodding landscape is changing: "From everything we can see at the magazine there are plenty of pre-'48 hot rods being built but there's no denying that the building of these early hot rods are slowing down as the later models are increasing. If one is to look at the total numbers of the later cars (and trucks) participating at all events the numbers would seem to be growing."

Chapter 25

1. David Dudley, "American Beauties." *AARP Magazine*. April/May 2012, p. 46–49, 77.

2. I am referring to a rise in the proportion of the rodder population that is elderly. The impacts associated with aging are considerable. Issues directly impacting levels of participation include health status, pension, general loss of wrenching productivity due to mobility issues, and the loss of friends involved in rodding.

3. The H.A.M.B. ("Hokey Ass Message Board") is at http://www.jalopyjournal.com/forum/. On July 10, 2016, according to the site, there were 920 members and 2,585 guests online. The Bangshift Forum is at http://www.bangshift.com/forum/. At midday on July 10, 2106, there were 221 users on the forum.

4. For a good sense of the brain trust of online forums, see "Motoguy H.A.M.B post #1 [July 12, 2016], [Technical]Who works on Roto Hydramatics?, specifically the comment by "bobss396" [#11] (The oldest transmission shop in town was the ticket for old hydros ... but all the guys who knew their shit are long gone. The going price was $1,200 by me 15 years ago. I sealed one in my '64 Olds when I was 18. It worked fairly well considering how long I drove it when it was pissing fluid like a race horse.). Also the following comment by "poncho60 [#12] (If it's a stock 63 Star Chief, it does not have the Roto [Slim Jim] ... it has the dual coupling hydramatic.... Known as the Jetaway, Super Hydro, etc. The dual coupling trans was not the one modified by B&M and others for drag racing. Don't know why there is still so much misinfo out there regarding the dual coupling vs dual range. You really can't mod the dual coupling to increase performance.).

5. See "Milestone Custom Cars: Bob Hirohata '51 Mercury" a discussion in "Traditional Customs" in Jalopyjournal.com started by Rikster, August 28, 2005. http://www.jalopyjournal.com/forum/threads/milestone-custom-cars-bob-hirohata-51-mercury.68232/. Also see Rik Hoving's "Custom Car Photo Archive: '50s Hirohata Merc Photos" at https://public.fotki.com/Rikster/11_car_photos/beautiful_custom_cars/barris-1/hirohata_mercury-1/50s_photos/.

6. Pers. Comm., May 31, 2016.

7. Pers. Comm., May 9, 2016.

8. H.A.M.B. post #77[February 21, 2017], "So ... at what age did you have to give up building hot rods?" by v8deuce [Randy Haugen].

9. Pers. Comm., April 9, 2016.

Selected Bibliography

Alig, J., and Kilmer, S. (2012). *East vs. West Showdown: Rods, Customs & Rails.* Forest Lake, MN. CarTech Press.

Borg, K. (2007). *Auto Mechanics: Technology and Expertise in Twentieth-Century America.* Baltimore, MD: The Johns Hopkins University Press.

Crawford, M. (2009). *Shop Class as Soulcraft: An Inquiry into the Value of Work.* New York, NY: The Penguin Press.

Drake, A. (2008). *The Age of Hot Rods: Essays on Rods, Custom Cars and Their Drivers from the 1950s to Today.* Jefferson, NC: McFarland.

Furman, M. (2004). *Automobiles of the Chrome Era 1946-1950.* New York: Harry N. Abrams.

Ganahl, P. (2014). *Hot Rod Gallery: A Nostalgic Look at Hot Rodding's Golden Years: 1930-1960.* Forest Lake, MN: CarTech.

Jetter, R. (2004). *Bangin' Gears & Bustin' Heads.* Baltimore, MD: PublishAmerica.

_____. (2008). *Recollections, Regrets, & Random Acts.* Baltimore, MD: PublishAmerica.

Lucsko, D. (2008). *The Business of Speed: The Hot Rod Industry in America, 1915-1990.* Baltimore, MD: The Johns Hopkins University Press.

_____. (2016). *Junkyards, Gearheads, and Rust: Salvaging the Automotive Past.* Baltimore, MD: The Johns Hopkins University Press.

Parsons, J., Padilla, C., Arellano, J.E. (1999). *Low'n Slow: Lowriding in New Mexico.* Santa Fe, NM: Museum of New Mexico Press.

Starbird, D. and Bledsoe, B. (2006). *Darryl Starbird: The Bubble Top King Custom Car Creations.* Afton, OK: National Rod and Custom Hall of Fame.

Index

A&W drive-in 17
AACA (Antique Automobile Club of America) 56, 58, 89
AARP Magazine 181
Ackerman's (Rochester, NY) 70
Advanced Plating 59
affluenza 21, 190
Afton, Oklahoma 72, 74–75, 81, 190
Alamosa, CO 92, 95, 99
Alamosa Early Iron car show 96
Albuquerque, NM 122
Alig, Joe 194
All Stars 63
Alloway, Bobby 28–34, 135
Alloway's Hot Rod Shop 27
Allstate Vespa Piaggio 50
alternative fuels 9, 129
Amelia Island 52, 56
America's Most Beautiful Roadster Award 31
American dream 81, 167–168, 174
American Dreams, Lost and Found 25, 190
American Graffiti 118, 158, 180
American Lafrance 86
Americana 122
Americruise 133
anonymous styling 3, 6, 186
Anthony, Sebastian 192
Assholes Garage 42
Austin, Bob 101–109
authoritarianism 11
Auto Meter custom gauges 18
autocross 33, 59
Automobiles of the Chrome Era 57, 195
autonomous vehicles 10, 171–172, 176, 178, 193

baby boomers 34, 20, 53, 59, 121, 149, 171, 178–181, 194
back-halfed chassis 191
Bailey, Joe 29

BangShift 182
Bangshift Forum 194
Barnum, Ron 31
Barrett-Jackson auction 181
Barris, George 41, 47, 73, 79
Barris Brothers 35, 39
Barron's 167, 191
Baskerville, Gray 29
Beatnik Bandit 85
bench racing 183
Ben's Machine Shop 98
Bertilsson, Bo 193
Best, Jack 75
Best of Car Talk 192
Betz, Stan 37
Black Hole of Auto Metaphysics 192
Bledsoe, Brice 195
Bobby Alloway 27, 29, 33, 135
Bomgardner Award 56
Bondo 32, 39, 55–56, 116
Bonneville 120, 183
Boudette, Neal E. 192
Brajkovich, Brylen 85–92
Brass Era cars 59, 81
Brennan, Brian 190–191, 193
Brookville 67, 106, 107
Brookville Roadsters 67
Brownsberg, IN 63
bubble top 75–80, 85, 190, 195
Bucky's 103
Buehler, Gary 66, 184–185
Buffalo Auto Show 103
Bunkley, Nick 191
Burger, Gerry 193
Burn Foundation 52
Busutti, Rick 8

Caldwell, Bill 161
camaraderie 92, 101, 113, 131, 138, 143
Canadians 107
Car Craft Magazine 75–76, 78
Carlisle 5, 67, 85, 90–91, 156
Carr, Gary 73

Carroll's Restaurant (Rochester, NY) 71
Casale, Jerry 8
Cayuga Performance 18
Cecil the Diesel 83
Central New York region 5, 10
Chapouris, Pete 30, 86
Chase, Crystal 182
Chase, Steve 186
Cherry, Wayne 81
Chevy small block 11
Chisenhall, Jack 133
chop, three-inch 73, 107, 109
Christenson, Eric 98
Civic Stadium 72
Classic Instruments 138
classic trucks 7, 165
Classic Trucks Magazine 10, 16, 85, 183, 190
CNC 7, 98–99
Coast to Coast 11
Coddington, Boyd 30–32, 35, 79, 135
Colby, Sandra L. 193
Colorado University School of Medicine 94
Concours d'Elegance 53, 56, 58
cookie-cutter hot rods 162
coronal mass ejection 177
Cortland, NY 15
Cortland Antique Automobile Club 181
Corvette Museum 56
Cotter, Tom 193
craftsmanship 10–11, 20, 184
Cragars 121
Crawford, Matthew 193
cruises 72, 130
cruising 21, 49, 72, 88, 125, 127, 131, 142, 168–169, 180
Currie 9-inch Ford 16
Cushenbery, Bill 78
Custom Classic Trucks 10, 16, 82, 85, 183, 190

D'Agostino, John 190
Dakota Classic Cruisers 131
Davis, John 93–100
Day, Dick 78
Daytona 72, 128
Dean, James 123, 191
defiance of the contemporary 3, 5–7, 27
definition of a hot rod 66, 92
De Graaf, J. 190
demographic transition 4
DeRancy, Jack 47
Detroit Autorama 56, 85, 87–88, 91–92
Detroit Speed 33, 138
Devo 7
Die Cast Hall of Fame 77
Dipol, Gino 40
Dipple, B. 193
dirt racing 29, 60, 62, 127, 132, 180
Dixey, Jerry 153–159
Dog 'n Suds (Rochester, NY) 71
Don't touch my car guys 65
Doom of Zoom 177
dream cars 178, 193
drive what you build 8
driverless cars 4, 7, 171–174, 176–177, 184, 187, 189, 192
dry lakes cars 70
Dudley, David 181, 193
Dunkirk Speedway (Dunkirk, NY) 72

Earl Scheib 39
Early Irons Club 96
East Carbon, UT 36
East coast–style dragstrips 72
Effingham Frog Follies 145
El Monte 118
emission controls 10, 16, 121, 168, 191
Energy Independence and Security Act of 2007 9
entropy 7
environment 5, 8, 10–11, 21, 113, 137–138, 151, 158, 168–170, 172, 175, 178
era, end of 8–9, 184, 186
Erie Dragstrip 72
ethanol-blended gas 12, 169
Evans, Rich 162
experimental cars 79

Farnetti, Tony 167
fat arming 123
Fayetteville, AR 37
Fenical, Ken "Posies" 86
Fioto, Gary "Chopit" 45
flex-fuel 12, 167
Flowmaster 90
Fontinalis Partners 172

Foose, Chip 34, 79, 82–83, 88, 108–109
Forcasta 75
Ford in a Ford 109
Ford Motor Company 95
fracking-related supply boom 169
Freemark, Yonah 192
fuel costs 10, 169
fuel-injected engines 11
Fukushima nuclear disaster 176
Furman, Michael 195
futuristic cars 77, 79

Ganahl, Pat 193
gearhead 3–4, 12, 27, 128, 179, 184, 186
Gettysburg, SD 13, 16
Gilbert, Daniel 21, 190
Gilbert, John 10, 16
Gingerelli, Dain 193
Gold Spinner 56
Goodguys 5, 25–26, 28–29, 32–33, 48, 57, 59, 132, 134, 145, 154, 171, 183–184, 191
Goodwill, Bill 24, 29, 32–33, 48, 183
Gosson, Scotty 193
Grand National Roadster Show 39–40, 42, 45–46, 78, 145, 147
Greens, Jessie 34
Greenwich 53
Gross, Ken 193
Gunns, Tommy 90

H.A.M.B. 143–144, 182, 186, 194
Harley 70–71, 149, 154
Hartford Show 103
Haugen, Randy 186, 194
Haukenes, Eline 161
Hertz Rental Cars 121
HFO-1234yf 138
Hines, Bill 39
Hirohata, Bob 183, 194
hobby 5–6
Holdaway, Ken 139–152
Hollenbeck, Darryl 190
hot rod (definition) 4
hot rod culture 4, 179, 181
Hot Rod Magazine 29, 68, 75, 101, 103, 133, 153
Hot Rod Power Tour 127–128, 133, 157, 169
Hot Tuner Nights 59
Hot Wheels diecast toy 60, 76–77, 132
House of Kolors 52
Houston Classic 53
Hoving, Rick 190, 194
Hrehovcsik, Greg 93, 98–99
Hudson Select 103
Huff, Bo 35–47
Hunter, Sarah 173

Huntington Beach Auto Body 162
hybrid technology 10–11, 70, 130, 167–168, 175
Hyundai 48

IMCA 64
Inland Empire 19
ISCA 30–31, 56, 60, 190

Jacobs, Chris 42
JC Whitney 94, 96, 130
Jeffries, Dean 79
Jerome, John 13, 20, 189
Jetter, Roger 118, 123, 190–191, 195
Jim Ray Chevrolet 37, 44
Jimmy's Bar (Warren, MI) 127
Joey Chitwood Thrill Show 61
Johnson, Alan 34, 135
Johnson, Eastman 8–9, 189
Johnson, Walt 139–152

King, Jonnie 190
King, Stephen 86
kit car 52, 106
KKOA 73, 139, 145, 155, 160
Klein, Gabe 172
Kustom Kulture 35, 73, 139, 190

Lacquer 37, 39, 105–106, 136, 160
Langston, Nile 97
Lassa, Todd 193
Lead Sled Spectacular 139
Leopold, Aldo 5–6
Little Coffin 76
Little York Car Show (Little York, NY) 19
Livy, Titus 11, 189
Loh, Edward 192
Long Haul Power Tours 128
Love, Rick 132
lowriders 47, 171
Lower Law 39
LS series engines 11, 16, 18, 59
Lutz, Bob 165–166, 191

Maaco 39
Magliozzi, Tom and Ray 175, 192
Manning, Ronnie 144
Martin, Colby 191–192
Masterson, Brad 39
Matchbox cars 85
matching numbers 111, 116
Mattel 76
Mayes, Alan 193
McLeod, John 138
McMullen, Tom 31, 157
Meadors, Gary 25, 88, 183
Meadow Brooke 53

Index

mechanical cluelessness 4, 27, 177, 187, 193
message in a bottle 46, 81, 117, 138, 151, 162
Messler, Gary 130
metal flake 36–37, 50, 60, 79
midgets 72
Milwaukee Art Museum 8–9
Minnesota Street Rod Association 5
Mint 400 72
Model A 38, 60, 64, 67–68, 94–95, 103, 112–113, 160–161, 175, 184–185, 192
Model T 42, 59, 82, 101, 113, 154, 176
Monogram 75–76, 79
Monroe County Fairgrounds (Rush, NY) 61, 72
Monster Garage 162
Monta Vista, CO 97
Mothersbaugh, Mark 8
Motion Supercars 111
Motoguy H.A.M.B. post 194
Murphy's Law 177
Murray, Larry 31
muscle car 5–7, 11, 21, 34, 38, 65–66, 86, 110–111, 116–117, 121–122, 132, 134, 137, 149, 154, 168–169, 180–181
Muscle Car City 110
Musk, Elon 192

NASCAR 63, 165
National Roadster Show 39, 40, 42, 45, 78, 145, 147
National Rod & Custom Car Hall of Fame Museum 73–74, 190, 195
National Street Rod Association 5, 59, 153
Naylor, T.H. 190
NCRS 56
NHRA Jr. Championships 129
Niagara Raceway Park (Niagara Falls, NY) 72
Nick Matranga Merc 42
Nikki 111
nostalgia 9, 21, 107, 112, 125, 162, 180–181
NSRA 59, 134, 145, 153–154, 156–157, 171
NSRA's 29 Below program 134

Offenhauser 191
Ol' Skool Rodz 40
Old School 9–10, 61, 119, 171, 185
The Old Stagecoach 8–9
Olin, IA 101
Omaha Auto Auction 129
Ortman, Jennifer M. 193
Osborne, Lee 61–72

OSHA 96
oval tracks 72
Overhaulin' 42

Pace, Clyde 95
Painless 18
Parker, Garland 97
passion 7, 11, 27, 112, 131, 138, 185–186
peak oil 8
Pebble Beach 46–47, 53, 56, 86
Pete & Jake's Hot Rod Parts 30
Petersen, Rick 127
Petersen Automotive Museum 179
Petty, Rod 118–126
Phaeton 31, 101, 105–106, 108
Plouffe, David 172
plug and play 11, 20
poncho60 H.A.M.B. post 194
Popular Mechanics 96
Porter Ranch gas leak 176
Posies Rods and Customs 86
Power Tour 113–114, 127–128, 133, 157, 169, 183
PPG Hot Wheels Anti-Freeze Green candy 60
Predicta 76, 78–79
President's Cup 56
Prius 69–70, 175, 192
prototypes 79–80
Puppycrusher 85, 88–90

R134a 138
rat rod 4, 42–46, 59–60, 65–66, 73, 85, 119, 128–129, 162
Rat's Glass 31, 190
Rebel Without a Cause 191
Renaissance Delivery 31
Revell 79
Rick's Place 108
Rickshaw Car Club 96
Ridler 28, 30–31, 34, 60, 87
Right Coast Association 171
Risch, Carl 48
Road Gents 96
Robert Pass (Passport Transport) 86
Robles, Stan 39, 44
Rochester War Memorial 72
Rockabilly 35–36, 41, 44–45, 171
Rockabilly Hound Dog 190
Rod and Custom Hall of Fame 32, 74, 195
The Rodder's Journal 193
Rodriguez, Johnny 96
Roedar, Mike 145
Roetman, George 127–131
Rolling Bones Hot Rod Shop 182
Ross, George 48–60
Roth, Ed 78, 85

Round River Journals 5
Route 66 45, 78, 122
Russell, Charlie 178, 193

St. Antoine, Arthur 192
Salina, KS 139, 145, 160
San Bernardino, CA 122
Santa Casho, CO 94
Santa Monica, CA 122
Savannah Drag Strip (S. Butler, NY) 72
Scott, Gene 121
Scottsville Road (Rochester, NY) 72
SEMA 45, 90, 132, 137–138, 156, 172, 191
Shangri-La Speedway (Owego, NY) 63
Shimmering hues 80
Silicon Valley entrepreneurs 154
Smith, "Speedy" Bill 90
So-Cal Speed Shop 30
Smith, Tex 121
SODAC'S 130–131
Sorokanich, Bob 192
South Dakota 13, 15, 127, 129–131
Southard, Andy 193
Speartech Fuel Injection Systems 17
Speedy Bill 90
Spencer Speedway (Williamson, NY) 64, 72
Spencerport, NY 72
Sport Compact Division 129
Springer, Ron 190
sprint car racing 63
Stallone, Sly 86
Starbird, Darryl 32, 73–84, 146, 190, 195
state of the industry 137
Stella 44
Stray Cat Customs 144
Street Rod Nationals 49, 153–154
Street Rodder 30, 43, 45, 132, 138, 153, 155, 157–158, 183, 191, 193
street rods 4, 7, 10, 28–29, 33–34, 132, 137, 49, 156, 168, 175, 180
StreetScene 3–4, 184, 189, 191
Suede Palace 46
SureFit kit 135
survivor cars 59
Syracuse Nationals 4, 65, 161

TCI 4 link suspension 16
Terkel, Studs 25, 190
Tesla 175, 192
Thompson's Station, TN 48
3D printers 7, 183
Thunder Valley Dragways (Parker, SD) 129

Titus, Jerry 73
Top Flight (Bloomington, IN) 56
Town Line Road (Spencer, NY) 103
trailer queens 8
Trails Plowed Under 178, 193
Trepanier, Troy 34
Treworgy, Rick 110–117
Tri-5 28–29, 34, 50, 119
Triple Cities Street Rods 132
Triple Crown 56
Tubbs, Randy 19
Tucker, Kyle 33, 138
Turner, Ray 61–63, 69
Television build shows 20

Uber 172
Udy, Jason 192
Upstate New York 8, 15, 181, 183
urethane 105

USAC 63–64
v8deuce H.A.M.B. post 186, 194
The Valley Courier 96
Van Sickle, Neal 93–95, 98–99
Van's Machine Shop 93, 95–98
VanZile, Ray 48
vendor events 72
venture capitalist 172–173, 176–177
Vermillion, SD 127, 129
Vietnam 63, 118
Vincent, Peter 193
Vintage Air 132–134, 137–138, 158
Viva Las Vegas 45, 190
Vlasic, Bill 192
Vogele, Tom 3, 155–157

Wann, D. 190
Warshawsky 94, 96
Watkins Glen Eco-Challenge 191

Weiand manifolds 94
Weikel, Dan 192
Wentz, Dakota 73, 75, 146, 190
West Whitlock State Park, SD 13
Western Iowa Tech Community College 128
Wiggins, Ken 107
Winfield, Gene 20, 79, 160–162, 149, 179
Woolsey, R. James 193
World of Outlaws 64
World of Wheels (Detroit) 91
World War II 4, 160, 179
WyoTech 44

Yenko 111
York, Pennsylvania 107

Zip, Darrell 79
Zipcar 172
Zubal, Freddy 39

www.ingramcontent.com/pod-product-compliance
Ingram Content Group UK Ltd.
Pitfield, Milton Keynes, MK11 3LW, UK
UKHW050526150426
5217IPUK00026B/1817